S0-CCO-699

GLANDS—
OUR INVISIBLE GUARDIANS

▽ ▽ ▽

Glands—
Our Invisible Guardians

BY

M. W. KAPP, M.D.

ROSICRUCIAN LIBRARY
VOLUME XVIII

SUPREME GRAND LODGE OF AMORC
Printing and Publishing Department
San Jose, California

Second Edition
December, 1941

Third Edition
June, 1947

Fourth Edition
March, 1949

Fifth Edition
August, 1954

THE ROSICRUCIAN PRESS, LTD.
SAN JOSE, CALIFORNIA

Sixth Edition—July, 1958, with additions
and revisions by Stanley K. Clark, M.D., C.M., F.R.C.
Seventh Edition—April, 1960
Eighth Edition—October, 1962
Ninth Edition—December, 1964
Tenth Edition—November, 1966
Eleventh Edition—June, 1968
Twelfth Edition—January, 1972
Thirteenth Edition—August, 1973
Fourteenth Edition—August, 1975
Fifteenth Edition—June, 1977
Sixteenth Edition—April, 1979

Printed and Bound in U.S.A. by
Kingsport Press, Inc.
Kingsport, Tennessee

DEDICATION

▽

To The

Men and Women

throughout the world who are seeking to understand the physical, mental, and spiritual urges of their beings,

This Book Is Dedicated

with the hope that it will serve them in their noble purposes.

1939

The Rosicrucian Library

VOLUME

I Rosicrucian Questions and Answers with Complete History of the Order
II Rosicrucian Principles for the Home and Business
III The Mystical Life of Jesus
IV The Secret Doctrines of Jesus
V Unto Thee I Grant (Secret Teachings of Tibet)
VI A Thousand Years of Yesterdays (A Revelation of Reincarnation)
VII Self Mastery and Fate with the Cycles of Life (A Vocational Guide)
VIII Rosicrucian Manual
IX Mystics at Prayer
X Behold the Sign (A Book of Ancient Symbolism)
XI Mansions of the Soul (A Cosmic Conception)
XII Lemuria—The Lost Continent of the Pacific
XIII The Technique of the Master
XIV The Symbolic Prophecy of the Great Pyramid
XV The Book of Jasher
XVII Mental Poisoning
XVIII Glands—Our Invisible Guardians
XXI What to Eat—And When
XXII The Sanctuary of Self
XXIII Sepher Yezirah
XXVI The Conscious Interlude
XXVII Essays of a Modern Mystic
XXVIII Cosmic Mission Fulfilled
XXIX Whisperings of Self
XXX Herbalism Through the Ages
XXXI Egypt's Ancient Heritage
XXXII Yesterday Has Much to Tell
XXXIII The Eternal Fruits of Knowledge
XXXIV Cares That Infest
XXXV Mental Alchemy

(Other volumes will be added from time to time. Write for complete catalogue.)

CONTENTS

∇

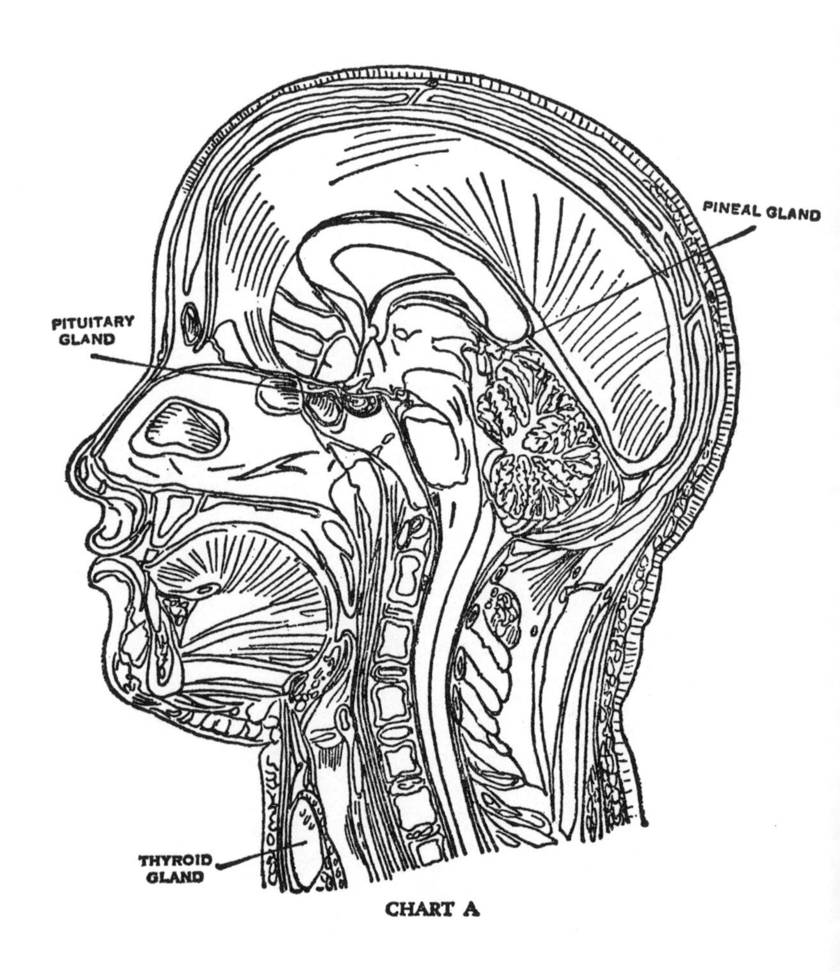

CHART A

INTRODUCTION

Every activity of Nature is within Cosmic Laws. There are no miracles. The author and the publishers hope to bring to the lay mind, in simple language, many of the truths now known to science and philosophy and thus pave the way for more truths and knowledge that may be woven into Wisdom.

Surely every thinking mind will realize the need for truths and knowledge. It is only through these that wisdom may be attained.

Falsehood, superstition and ignorance can never lead us to wisdom, nor to moral, mental and spiritual beauty.

There is no standing still in our evolvement. We either progress or retrogress. This law is Nature's edict, not man's.

Man's evolving as far as he has is due to Cosmic intuition or poetic genius that has driven him ever on, and mostly upward. All through the ages he has been seeking more light, more power, more happiness, and more love.

Science is opening the doors to a more positive advancement for man. It is proving that life's processes for man lie almost wholly within himself and are amenable to control and upbuilding.

In the past man has been taught by exhortations, by affirmation, by superstition, but now we have teaching by proved and demonstrated facts.

Science now knows that the building and evolvement forces within man operate almost wholly through the glandular structures of the body. We hope to bring the known facts of the glandular activities to the reader so that he will understand them and thus be able to live a fuller and more beautiful life.

This volume is dedicated to all who seek more truth, knowledge and understanding.

M. W. Kapp, M.D.
San Jose, California

FOREWORD

The one great unanswered Question is:—"WHAT IS LIFE?"

We are not foolish enough to try to answer that profound question.

All we know is what lies within our consciousness. Our consciousness tells us that each of us is a "Being" or in a state of "being." If I am a being now, I must have been in a state of being before I arrived on this plane of being, or I must have been created from nothing.

While we may not know what life is, we may study the manifestations of life and profit thereby.

From the oldest of philosophies and from modern science we learn that Man has four primitive urges.

The First Urge is for POWER and is manifested from the time the child begins to move about, and through development as a child, and an adult, and even to old age. In games, in fights, in social preferment, in politics, in finances—always seeking more power.

The Second Urge is for POSSESSIONS and is fostered from the time the child reaches out for a colored rattle on to the possession of estates and all forms of wealth and bodily comforts.

The Third Urge becomes the Love of Life and the Creative Urge. This is the differentiating force. It is the manifestation of the male and female. It includes the urge for parenthood, the love of humanity, the intellectual uplift.

The Fourth Urge is for Spiritual Uplift and understanding of moral beauty and the advancement of human relations and the "God Within."

If the reader will constantly keep in mind the Four Urges, he will better understand the building and driving forces of the ductless glands. The story of the ductless glands, though a comparatively recent study, is a very vital one.

These urges are impelling and are only slightly influenced by the voluntary mind forces. A calm nerve or mental state will aid normal activity. A worried or harried state of mind retards the glandular activities. The emotions of the body come from the endocrine or gland activities.

In the study of these forces and urges within Man, one must always keep in mind the Law of Action and Re-Action, or the Law of Cause and Effect.

My knowledge and understanding of the ductless glands and their activities has helped me very much in my 43 years of active medical practice. I wish to express my appreciation of the debt I owe to the work of such men as Sajous, Crile, Cannon, Loeb, Lorand, Berman, Bandler, Millikan, Soddy and many others. These men have brought us a vast heritage.

To play the game of Life well, one needs to know the laws and rules of the game.

Alexis Carrel in his wonderful book *Man the Unknown* calls for scientists, philosophers, economists and teachers to awaken, and to give freely of their knowledge and efforts to uplift mankind. His call has inspired us to bring out this story of the ductless glands and man's evolvement.

M. W. KAPP, M.D.

CHAPTER I

Divine Alchemy

By H. SPENCER LEWIS, Ph.D.*

I AM very greatly pleased to be able to offer to our members and our friends this unusual manuscript dealing with the glands of the human body. Not only does the matter deal exhaustively with the subject, but I believe it only fair to state that the author of this manuscript, Dr. M. W. Kapp, has had a very remarkable career for many years in Santa Clara Valley and Central California, where his confreres and brethren of the medical fraternity hold him in the highest esteem for his efficiency, his unusual diagnostic abilities, and his keen insight into the invisible mechanism of the human body and its functionings.

Admittedly Dr. Kapp has had unusual success with his patients, and admittedly he has had a very definite viewpoint regarding health and illness, and this manuscript of his dealing with the glands of the human body and their functioning reveals why he has been able to assist his patients in receiving the natural, curative operations of nature. He has given not only nature, but the divine forces throughout the human body, a greater opportunity to do their normal, creative work. He has been able to point out to his patients why and how they have been doing those things, or thinking those things, or permitting those things in their lives that interfered with this normal, natural, divine functioning in their bodies.

Originally, Dr. Kapp called his manuscript *A Story of How the Lives of Human Beings Are Controlled.* In a

* Written in 1939

large sense his title was indicative of what he has really learned and observed. While he has not been a member of the Rosicrucian fraternity, and he has not received the secret Rosicrucian instructions regarding the functioning of the various glands throughout the body, he has been, by nature and by insight, mystically inclined to such an extent that he has been able to observe, by carefully studying the lives and especially the abnormal conditions of his patients, how these glands have actually "controlled" the normal functioning of the entire body. The Rosicrucians have maintained for centuries that these glands act more like *guardians* of the lives of human beings than as *controllers* and yet it must be admitted that a true guardian is also a controller.

When we stop to realize that man in his earthly existence is functioning as a dual being, and that there is a spiritual self within a physical body, and that the spiritual self is there for the purpose of giving man intellectually a sense not only of divine wisdom and divine mastership over earthly conditions, but to guard and control the perfect operation of the physical body, we must realize that there must be also some means of exchange or communication between the spiritual self and the physical self. In other words, there must be some places or points within the human body where the spiritual power, self, and intelligence can transmute its power, authority, and control into the grosser elements of nerve energy, blood, vitality, and human mechanism so that the higher, finer, almost intangible and imperceptible forces of the divine self may be brought down to a rate of vibrations and a form of power crude enough, or material enough, to function through the flesh and bones and other material, chemical elements that constitute the body of man.

The glands have been found to be these intercommunicating instruments, these transformers, or transmuters between the spiritual, divine, Cosmic self and the grosser, earthly and physical self. They bring about within man a *divine alchemy*. For many centuries the most eminent mystical scientists, who made a very serious study of the rhythmic, synchronous functionings of both the divine and physical self in man's body, believed that the pineal and pituitary bodies, now known to be glands, were the only actual physical, material organs for such transmutation of a higher force and energy into a more material force. On the other hand, there were those who believed that the solar plexus was a gland of great importance, having the function of interpreting and transmuting the higher, inspirational, Cosmic, or spiritual emotions within man into the grosser, material, and emotional reactions. For a century or more the solar plexus was somewhat worshipped and adored as the seat and soul of all of man's higher activities. But when it was discovered that the spiritual element within man is to be found in every living cell of every part of bone, tissue, and blood, and that the soul and emotional nature of man are not located in one organ or one part of the body, it became necessary to study man's physical anatomy more carefully. Then the many other glands were discovered and given proper attention.

In this book, which we are happy to present to our members and friends, Dr. Kapp has explained, as a medical man and as a medical scientist would explain in psychic and mystical terms, the location, functioning, and purpose of each one of these main glands. He has done so in a manner that is not only consistent with what is contained in the Rosicrucian teachings, but is free from the more or less limited technical phrases and definitions of these teachings. Therefore, medical men, scientists, and laymen alike who have not been versed in the Rosicrucian terminology and principles can also understand the importance of these glands and their functioning and the other necessary conditions for keeping the body normal in every possible way and permit these glands to perform their divine purpose without earthly, material, physical interference.

Speaking of the emotional centers of man's body again, we have found, as have scientists and medical men, that the spleen is just as reactive and just as demonstrative of the emotional functions of man's mental, psychic, spiritual, and physical existence as is the solar plexus. This, too, was discovered many centuries ago, and for that reason many popular phrases were invented by the more or less ignorant laymen whereby they expressed the idea that one who was despondent or unhappy or cranky was manifesting a bad spleen. But it is also true that no part of man's spiritual and physical composition can be out of order or out of harmony with the Cosmic rhythm or with the Cosmic flow of vibrations without man's emotions reacting and manifesting the inharmonious attunement.

From many mystical and spiritual points of view the pituitary and pineal bodies or glands may be quite important in certain so-called "psychic" reactions. No one knows better than do the Rosicrucians that these two glands or bodies should be given careful thought in connection with many forms of development of the latent spiritual, or Cosmic, abilities of the human being. But then again there is the thyroid gland which, while it does have a considerable importance in connection with the development and growth of the physical human body, and from the physical, medical point of view may be closely related with certain forms of malignant or toxic conditions that are subnormal or abnormal, on the other hand, is important in certain forms and degrees of psychic or spiritual development.

But all of the glands have some relation not only to the emotions within the average human being, but also to the mental tendencies. We are now learning that certain types of criminals are unquestionably victims of certain gland conditions and should be classified as *gland criminals.* The endocrines offer an opportunity for criminologists to definitely foretell the tendencies of a criminal nature on the part of those who have just passed through the adolescent stage and are approaching adulthood.

It is not necessary for every individual to become mystically inclined or to be given to the study and reading of mystical, spiritual, or religious subjects in order to be benefited by a very careful study of the glands within the human body. Undoubtedly certain forms of extreme religious fanaticism are due to an abnormal or subnormal development, or an atrophied condition of certain glands. But do not allow this to give you the impression that to be devoutly religious, or even to put religion above all other things, is necessarily an indication that you are overdeveloped in regard to some of your glands. The atheist, like some foreign and American medical writers, would have us believe that an enthusiastic belief in, or an enthusiastic adherence to, any doctrines or practices indicates a subnormal or abnormal mental and glandular condition. This is not true. The truly normal, natural person is one who does express and manifest certain very definite, enthusiastic principles, ideals, and desires, and the really abnormal or subnormal person is one who "takes life as he finds it" and who from day to day finds no ecstasy, no joy, no happiness, no enthusiasm in any one thing that interests a part or all of human civilization.

That those who are enthusiastically inclined toward mysticism and psychic matters may be abnormal is true only if we take a cross section of the entire world population and make comparisons, because thereby we find that those who are enthusiastically inclined in such subjects are in the minority. But should we assume that the minority in any case is subnormal or abnormal? The line between perfect sanity and the slightest degree of insanity is so flexible, intangible, and indefinite that no one, not even the greatest psychiatrist, can attempt to define it and establish it. It has been facetiously said that all of us—meaning you and me and the rest of the world—are insane on some subjects. By that is meant that a majority of us are more enthusiastic, react more easily to some ideas, some ideals, and interests than do other human beings. But this is not an indication of an abnormal attitude of mind, or an insane attitude or a faulty development. It is simply a manifestation of the complexity of human nature, and of human emotions, which complexity makes human existence interesting and gives us the manifold manifestations of art and the creative abilities and the beauties of man-made things of a material or spiritual nature.

Even our human countenances, our human attractiveness, and most essentially that intangible something called *human personality* or *personal magnetism,* are the result of the normal and proper functioning of the glands. And that which attracts one person to another is something more than the mere definiteness of the handclasp or the deliberateness of the smile, or the wiles of the pleasant words that are spoken.

By knowing our glands and how they function, and by knowing how to live properly, which includes eating, drinking, and breathing properly as well as thinking properly, we can permit these glands to do their very best, and give us every advantage of their divine functioning. Dr. Kapp has very beautifully outlined these ideas in his manuscript, and I especially urge each reader to pay very strict attention to the first twenty or thirty pages of this book wherein many very excellent ideas are presented for the first time by a man who is above everything else a very strict and careful medical practitioner of many years' experience, and secondly, a keen and excellent student of human nature. With this introductory chapter, therefore, I recommend this book to our members and friends and their friends everywhere. Many books have been published recently dealing with the various glands of the body, and some even dealing with the effect of these glands upon the human personality, but most of them have been too technical, too much like reading a book on physiology or anatomy, and have missed or entirely ignored or negated the divine and Cosmic functioning of the glands and the spiritual side of nature.

I am glad, therefore, that one physician of truly scientific training, and especially one living so close to us here in this beautiful valley, and so greatly in sympathy with the work of the Rosicrucian Order, its research departments, and its Rose-Croix Research Institute and Sanitarium,* is able to add to the bibliography and literature of the Order a manuscript and book that will undoubtedly remain in its archives for many centuries to come.

* No longer in existence.

CHAPTER II

The Endocrines and Their Hormones

MAN WILL evolve when he no longer craves the bestial, the barbarous, the more physical, gratifications. When the time comes that he feels the spiritual urge and intellectual enlightenment which lead to feelings of exaltation and elevation (which must and will come to all men at some stage in their evolution), then there comes a quickening of all the moral senses and a consciousness of immortality and goodness.

The superiority of really great or illuminated soul-personalities, or egos, lies in (1) intellectual acuteness, (2) moral elevation, (3) all-embracing optimism, (4) the sense of immortality—or to express it more tersely, Cosmic Consciousness, or the more familiar phrase, realization of "the kingdom of God within us." Cosmic sense will give us the power over good and evil.

Man has been a long time evolving from his lower states of being. There is still a long upward climb ahead of him. It may be that we have all eternity to evolve in, and the slow method may be the best, but when one glimpses the beauty ahead—even for a few illumined moments—one becomes eager to evolve rapidly to a higher plane. One cannot stay forever in the state of self-consciousness alone. One must eventually evolve into the Cosmic Consciousness.

Our egos or soul-personalities (or whatever you choose to call that which we are) must have the physical body to function or perform in. This can be as pure as the mental or spiritual. We doubt, however, if a clean soul-personality can remain in a bestially inclined body. To keep the building forces of the body and soul clean and in normal activity is our great hope and aim. The building or constructive power of man comes through the endocrines and their hormones. The endocrines are the glands, mostly ductless, that secrete and distribute the substances called *hormones* which control our constructive powers, both physical and mental.

The story of the ductless glands or endocrines and their hormones is the story of the human race in its evolvement and unfolding. This story is just in the telling and cannot be all told until man reaches perfection or complete evolvement. We hope to bring to the lay mind some knowledge and help in the constructive processes of everyday life.

The main endocrines or ductless glands of the body are:—the pineal, pituitary, thyroid, adrenals, gonads (sex glands), and spleen. A few glands which have ducts also secrete and distribute hormone substances. They are the liver, pancreas, kidneys, salivary and lymphatic glands. We are also learning lately that the vitamins that are so vital to the nourishment and activity of the body are built through the glands just mentioned.

A serious study of the endocrines and their hormones began about fifty years ago, although about 170 years ago a French savant, Theophile de Bordeu, made some study of the subject for a few years. Forty-five years ago, Brown-Sequard made a life rejuvenating elixir, an extract of the testicle injected under the skin of the patient. Well do I remember a trip to the slaughterhouse with the old family doctor under whom I was studying medicine, and the selecting of a ram's testicle. From it we made the "elixir" and then made injections under the skins of three old men. The old men did not rejuvenate. The Brown-Sequard furore gave an impetus to more careful study and experimentation, the results of which have greatly benefited humanity.

The human body is built and vitalized by definite chemical processes, well described by science. We do not yet know all the processes of construction, but we may profit greatly by what has already been discovered.

Life forms are an expression of consciousness. As cells evolved and various consciousnesses became active, it became necessary that some control of conscious activity be formed; therefore, centers of control developed.

The cells needed control of light reaction, of pigment, brain growth and sex ripening, and so there became a center for all this—the pineal gland. Also there developed the need of nourishment and a center of control for the body building and proper food intake; so evolved the pituitary gland, which guides the senses of taste and smell by which we select proper food and reject the unwholesome. The size of our bodies and mental power is controlled by the pituitary gland. The cells of the forming body needed iron, phosphorus and arsenic, and to meet this need came the thyroid gland. All this evolving needed energy control, and a rhythmic movement producer; so the adrenals were formed.

The single cell was immortal and had power to renew itself perpetually, but when the cells began to work together they lost the power of everlasting life and had to resort to reproduction and general creative power. To produce this center for reproduction and creative power, the sex glands (gonads) were formed.

The growth of the young individual needed care, with a check on some of the glands of later evolvement, especially the sex glands, and there was created the thymus gland.

A storing place for food, a renovating system and a circulating and distributing system became necessary; so the liver evolved and the blood circulatory system and the lymphatic ducts and glands developed.

Thus Nature, or the Creative Force, built according to law and order with an astounding intelligence. No wonder the ancients said God was "Spirit," for this subtle force that builds is powerful and yet invisible. A man who studies deeply this force becomes unconsciously religious.

These glands must work in harmony, and to do so they needed methods of intercommunication. The chemical intercommunication is the oldest, we think, but soon another method evolved. This we call *nerve action*. We doubt if, even now, we know all the processes of conscious relation of the glands and body. There may be subtle vibrations not detected as yet. When we have learned the method of building and the relation of the glands of the body, then we shall be masters of our whole consciousness.

We had long supposed that the brain was the source of energy and the place from which the will acted, but now we know it is only the depository of memories, communicated to it by some process that leaves its records somewhat as the phonograph needle leaves its record.

The vegetative system (the gland system) is the seat of the impulses of man. Love is not a result of the thinking mind. It is an endocrine urge. A reasoning mind can help to stabilize love but the impulse does not come from conscious thinking. So it is with hates, fears, etc. War does not start by man's reasoning mind, but by the primitive urges from the endocrines. No reasoning mind will wish to kill. We need to build the reasoning mind, and to study the subconscious mind, so that we can direct the building forces constructively.

▽ ▽ ▽

CHAPTER III

LOCATION AND ACTION OF THE GLANDS AND ENDOCRINE SUBSTANCE

Pineal Gland

(SEE CHART A)

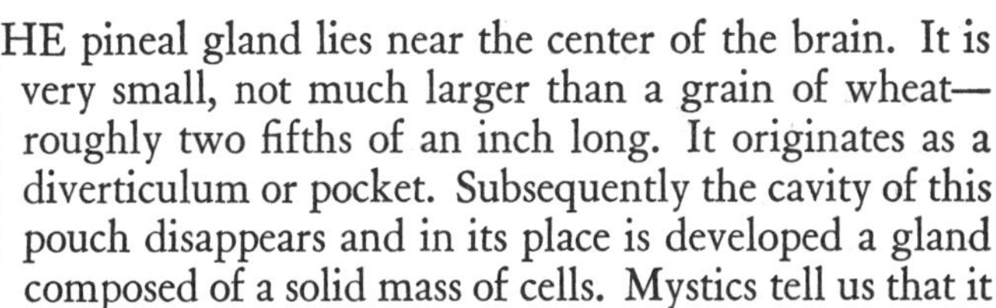

THE pineal gland lies near the center of the brain. It is very small, not much larger than a grain of wheat—roughly two fifths of an inch long. It originates as a diverticulum or pocket. Subsequently the cavity of this pouch disappears and in its place is developed a gland composed of a solid mass of cells. Mystics tell us that it is the bridge between the higher planes of consciousness and the physical plane of expression. Tradition tells us that it is the remnant of a *third eye* used by man in his earliest development. Descartes claimed that it was the seat of the soul. We know that there are cells in the pineal gland like those in the retina of the eye.

Life is built around lime salts and the X ray reveals that the pineal gland contains small grains of sand or salts. In disease the amount of sand in the gland is largely increased. In cases of tumors of the pineal gland there have been wonderful developments in sex and mental attainments and even in spiritual and loving affection. This gland helps in holding the sex in abeyance in childhood and aids in the ripening process of sex after puberty. In childhood it probably acts with the thymus gland. This gland seems to be the balance of control for the action of light upon the pigment of the skin. "It is the light within that reflects the light without." The pineal acts in conjunction with the adrenals in skin pigmentation. It also acts in conjunction with sex glands and brain.

Involutionary or regressive changes are said to commence in the human pineal about the seventh year of age.

Pituitary Gland

This gland is about the size of a pea and lies at the base of the brain behind the root of the nose in a little bony cup or cradle called the *sella turcica* or Turkish saddle. If this cradle is too small the development of the gland is retarded and the person is likely to be one of moral and intellectual inferiority. The condition of this cradle can be demonstrated by the X ray. The gland is composed of an anterior and posterior part. Each part has a distinct and separate origin, history, function and secretion. It is a sort of male-female combination. The anterior portion is a proliferation of the mouth area—the taste and smell sense area—and is considered the chief or master gland of the endocrine system. The posterior part is an outgrowth from the oldest part of the nervous system. The pituitary gland is often called the somatic brain for it seems to be the center of the subconscious action.

The pituitary gland can be traced from the most primitive form of life to man, and is the same in all planes. This gland and the salt of the blood we have brought all the way in our development from the sea to our present state. They dominated us then as they do now. The pituitary gland is a veteran of the ductless gland class and probably the most important in man's development. Its extirpation means death in a very short time. It has been called "Nature's Darling Treasure" because it is provided with a skull and is within a skull for its protection. Here we have an intimate meeting or mingling of the internal secretions and nerves, though little is known of its nerve supply. We doubt if much of its fineness of action is as yet understood. Experimentation has proved that the secretions from the anterior lobe of the gland stimulate growth of bone and connective tissue.

An extract of the posterior pituitary raises the blood pressure, stimulates plain muscle, such as the uterus, the intestines, the gall bladder, the urinary bladder, and the ureter. It checks kidney secretion and gives rise to an increase of sugar in the blood and urine. It also controls the salt content of the blood, which aids in the electrical conductivity of the system. It causes milk secretion of the breasts and, if given hypodermically, causes contraction of the uterus, aiding in expelling the child in childbirth. No other gland can take the place of the pituitary gland.

The anterior lobe has a balancing power over sex and creative force. The skeleton is dominated by the anterior pituitary and we have giantism or dwarfism depending upon the development and secretion of this lobe. Excess of anterior lobe secretion and lack of posterior lobe secretion makes the giant. An excess of anterior lobe secretion with an excess or increase of posterior lobe secretion makes the mental giant. This type is usually tall, bony, and strong. The subpituitary person is usually fat, lethargic, short, dull, sexually impotent, and is also likely to be an epileptic.

The pituitary controls the periodicity of sleep. An active pituitary means alertness and wakefulness. A tired or dulled pituitary means sleep or hibernation and general dullness. Feeding of the gland products is not yet as satisfactory in results as we might wish. There is something about the chemistry of it we do not yet know. Hypodermic medication seems to be rather active in results.

In pituitary excess the person is typically tall, lean (cannot fatten him), with a tendency to high blood pressure and sexual trends and great mental activity and initiative. He is sometimes irritable but has great endurance. The pituitary is the gland of continual effort.

When the cradle (sella turcica) for the pituitary is too small we get underdevelopment of intellectual and moral sense. Persons so afflicted are called pathological liars for they have no sense of truth. This condition also aids in producing morons.

The pituitary gland can be removed in animals without fatal results. The removal does however shorten the life span.

Among the effects of the removal of the pituitary are arrested growth, atrophy of the gonads and the accessory sex organs, suppression of the milk secretion, wasting or involution of the breasts, atrophy or wasting of the thyroid, the adrenals, and the parathyroids. Removal of the pituitary gland also gives rise to a lowered metabolic rate, hypoglycemia (less sugar in the blood), increased sensitivity to insulin, a lowering of the normal sugar content in the liver and muscles, a slowing down of the natural or normal activity of the animal, and the lowering of the animal's resistance to infection and shock.

The Thyroid Gland

This is located just below the larynx and is composed of two lobes lying on either side of the windpipe and connected by a "bridge" just below the Adam's Apple. It arises from the same tissue and almost from the same spot as the anterior lobe of the pituitary gland. It has a great controlling force in the growth of man's body and its sustaining power. The thyroid acts upon the growth of the inner and outer coverings of the body—skin, hair, glands, and mucous membranes. It is the builder of the nerves and brain tissues. It is essentially an energy-producing organ. It forms the greater part of the iodine or iron and phosphorus and arsenic of the system—iron for the general system and to aid in the electric energy and conductivity of the system; phosphorus for the nerve and brain centers; and arsenic for the skin.

In the lower forms of life the thyroid was a sex gland. It is now a link between the sex glands and the brain. It is so intimately connected with the sex glands or gonads that it is influenced by sex excitement, or menstruation or pregnancy. The thyroid is the gland that produced land animals and is very important in the evolution of forms, and also progression. The feeding of thyroid to a newt transforms it into a salamander—an air-breathing land animal. Tadpoles will not develop into frogs if their thyroids are cut away, but the evolvement of tadpoles into frogs can be hastened by feeding of thyroid substance.

The thyroid is also very necessary for the development and evolvement of a higher consciousness and psychic powers. Thyroxin, the active principle of the thyroid, is pure iodine. The thyroid secretions are the controllers of the speed of living. The more thyroid the faster one must live; that is, there is a greater intensity and capability of living when the thyroid is active. The cretin is a person with a great lack of thyroid secretion and is slow of movement, clumsy and awkward and stumbles when going up stairs. The cretin seems to have no soul or at least it does not develop until the thyroid develops. (By soul we mean a mental and spiritual condition.) Many cretins would become fine adults if they were fed thyroid extracts and given proper food and had good surroundings.

The pituitary gland keeps the salt proportion of the blood the same as that in the sea. The thyroid keeps the iodine proportion of the blood the same as the iodine of the sea. These proportions and elements have come all the way in life's evolvement from the life in the sea and still hold the balancing or stabilizing power of the body, physical, emotional and mental. The proportion of iodine is one drop to four and a half barrels of blood or of sea water.

The liver is the greatest user of iodine of any organ in the body. Without thyroid secretion there can be no physical unfoldment, no function or faculty, no complexity of thought, no learning and no responsive energies.

Excessive thyroid secretion as a disease is called *Exophthalmia.* An enlarged thyroid is called a goiter. There are three kinds: (a) simple goiters, which are unaccompanied by constitutional features; (b) goiters associated with a deficiency of the thyroid hormone (hypothyroidism); (c) goiters associated with an excess of the thyroid hormone (hyperthyroidism).

In hyperthyroid states where there is an excess of the thyroid hormone, a partial or subtotal removal of the gland is performed by the surgeon.

In hyperthyroidism the individual has a rapid pulse. He is nervous, has a muscular weakness, a fine involuntary tremor, protrusion of the eyeballs, and has an increased metabolic rate. There is a dissipation or wasting of the body tissues, especially the fat stores, often a rarefaction of the skeleton, disturbance of the carbohydrate metabolism, and an enlarged thymus.

A lack of thyroid secretion is called a strumous or cretin condition. Thyroid secretion increases gastric peristalsis and hastens all metabolic changes. Fine teeth usually indicate good thyroid activity. Dry scaly skin usually means inferior thyroid activity. When the activity of this gland is normal a person's ability to throw off poisons or infections is much greater than when the thyroid is defective. Thyroid is the great energizer. Its normal presence makes life worth while and its absence takes all the joys out of life.

Iodine is employed in the form of a salt (table salt) in many countries as a prophylaxis against goiter. This is necessary in certain parts of Canada, United States, New Zealand, England, and Switzerland.

Much of the original quantity of iodine in the soil has been leached away by rivers and streams and is now in the ocean, where the iodine percentage is high in comparison to the soil.

The Adrenal Glands

The adrenals are two in number. They are cocked-hat-shaped glands just over the kidneys, about as big as the end of one's finger. Like the pineal, pituitary and thyroid they have no ducts but are very vascular so there is much blood carried to them and the adrenal secretion is carried away in the blood stream to all the tissues of the body where it is used.

Each gland is composed of a cortex or outer portion and the medulla or inner portion, called the core. The cortex is derived from the same tissues that formed the sex glands. Vigorous or fighting animals have a large cortex or outer part of the adrenals as for instance the lion, tiger, buffalo, etc. Timid animals, such as the rabbit, have a small cortex. There is a very close relation between the adrenals and the gonads or sex glands. Before birth, disturbances such as tumors in the adrenals are supposed to be the cause of hermaphroditism (both sexes in one). Tumors, or disturbances in the adrenals, occurring after birth, cause premature sex development. Children of three or four years of age with such disturbances often appear as 14 or 15 with the characteristic sex conditions of that age: voice, hair growth, strong muscles and general virility. There is a close alliance between the brain cells, sex cells and adrenal cortex cells.

Adrenal secretion (called *adrenin* or *epinephrin*) energizes the muscles of the body and very especially the muscles of the circulatory system and the digestive tract. The adrenals seem to get their potential energy from the thyroid output of iodine. In excessive action of the thyroid we get excessive adrenin or epinephrin and, from that, excessive stimulation of the muscles and especially the muscles of the heart and the rest of the circulatory system. The heart will pound like a trip hammer in an exophthalmia case.

Adrenal secretion is often called the enzyme or enzymes (an enzyme is a digestive ferment). It supplies the base for saliva, pepsin, hydrochloric acid, liver, pancreatic and intestinal juices. The adrenal cortex acts upon the pigment cells of the body. In diseases of the cortex of the adrenals the skin becomes dark or pigmented or bronzed. This is called Addison's Disease.

Death follows quickly upon the removal of the adrenals. There have been found cases of apoplexy of the adrenals and the action was similar to the action of apoplexy of the brain. Adrenin or epinephrin is the product of the inner portion of the gland, the medulla. The secretion promptly raises blood pressure when injected into the system. Adrenal secretion tenses all the tissues of the body. The adrenal flow is brought to excess by pain, fear, excitement, rage, or any of the painful emotions. (Everyone has at some time in life felt the tenseness of body under great stress of emotions.)

The adrenals are glands of combat. They are the evolvement from the "Fight or Flight" age of man. Nature has evolved all the glands and tissues as they were needed, but as man developed individuality and with it selfishness and greed he arrived where he no longer ran with "the pack" to hunt. He wanted the best of things and separate from the rest of the individuals. To get the best of the other fellow he was always ready to "steal a march" or "put one over" on his fellow man, and this led to the "Fight or Flight" condition; that is, he had to fight with all the fury of his power or run away with all his power. It was "tooth and claw" or fleetness of foot that was the law of self-preservation or rather self-satisfaction. Excessive adrenal secretion was absolutely necessary for activating force in the "Fight or

Flight" condition. Excessive use of any of nature's forces will sooner or later pauperize those forces.

In spite of our boasted civilization we are still in the "Fight or Flight" era. Our fighting and running away may not be as crude as the cave man's but it is still as destructive to the body and soul of man. Our jealousies, hates, fears, struggles for wealth, power, position, our lusts, and superstitions all call upon the reserve supply of adrenal secretion —the fighting or energizing secretion—until the glands are exhausted and we wonder why so many die of heart disease (over heart action), Bright's Disease, diabetes, tuberculosis, cancer, and other diseases of diminished resistance. "Americanitis" is the result of our rapid living, or our refined "Fight or Flight" era. The system is under constant shock and the reserve energy is under call all the time. Shock is a blow to the emotions from fright, anger, worry, surgical operation or injury or some unpleasant experience. Under too long continued stress the ductless gland system or endocrine system ceases to function and the adrenals stop sending out the supply of epinephrin, which is the tensing fluid, and the heart slows, the blood vessels relax and the brain loses its blood and unconsciousness follows. The skin turns white because the blood is no longer driven to the surface and a general condition of collapse ensues, and if adrenal secretion is not supplied by artificial means or the latent forces within the person given a chance to recuperate, death will follow.

Unconsciousness has been the turning point for the recovery of many a patient. The emotions and fears were for the time blotted out and their inhibiting power was dispelled and the latent power within the endocrines had an opportunity to assert itself. Joy, hope, love, religious fervor and other inspiring emotions have freed the endocrines and given them an opportunity to act harmoniously and constructively. Life is too strenuous and, when we learn the things that the endocrinologists are trying to teach, we will play fair with the forces within us. When we live so fast, we are in a constant state of shock though we may not be in a faint or unconscious. We are in a lesser state of consciousness because of our tenseness than we would be if we knew how to be calm and harmonious in our being. Our tenseness of living causes us to suffer with fatigue, nervous exhaustion, sensitiveness to cold, loss of appetite, cold hands and feet, and a loss of the zest of life. We have mental inability, tendency to worry and weep and, as we said before, a general condition of "shock." All the glands are involved in shock but the adrenals are the specific glands of shock. The person with insufficient adrenal gland action is very apt to succumb to diphtheria and severe fevers.

Men who have an excessive supply of adrenal secretion (not exophthalmic) have great energy and unless well balanced by the pituitary are apt to be cruel and dominating and are often given to great sex excesses. The excessive-adrenal persons are among the politicians, bankers, captains of industry, and leaders of men. They are the men with terrific driving force. When there is a good pituitary balance with strong adrenal supply we get the great thinker and man of fine power, with gentleness and fine ideals.

Excessive adrenal supply in a woman makes her masculine and neutralizes her ovarian secretion. Such women become the leaders and command responsible positions. It is a safe bet that the first woman President of the United States will be of this type, or at least an anterior pituitary-adrenal centered woman. This type of woman is also prone to growth of hair on her face and body.

After all wars the nations involved are always in a state of shock. Men and women are not normal. Waves of crime and excesses follow every war. Women become more masculine from the constant shock of their adrenal endocrines.

The adrenal arousing of the men turns them to sex excesses for expression of force or even to crime. Advocates of war say we need wars to arouse the evolving forces within man. We do not believe it necessary to arouse the brutal to evoke the poetic and spiritual side of man.

More recently there has been developed from the cortex of the adrenals an extract which is having a wide use in the treatment of many diseases. Pains in muscles and joints are often dramatically affected and helped. The drug is employed in rheumatoid arthritis, in bronchial asthma, and in other allergic disorders, especially of the eyes and skin.

The Gonads (Sex Glands)

The gonads of the female are the ovaries, breasts, and uterus, and in the male they are the testicles, the penis and prostate gland. They are the generative or reproductive glands or sex endocrines. They are of external and internal secretion. The ovaries produce the ovum but they also produce an endocrine substance that vitalizes a woman and makes her feminine. The testes have as their external secretion semen which is the spermatozoon carrier and which is stored at the prostate gland. The internal secretion of the cortex of the testes is the male energizing force and that which makes him really male. It is the male endocrine.

Early life was reproduced and perpetuated by budding or fissure. That might have been enough to perpetuate life and mankind but the Great Consciousness seems to have created further for some purpose. Something more was needed to evolve individuality and differentiation. From some Great Wisdom came the evolvement of sex individuality and sex differentiation with characters of negative and positive expression. Sex urge has caused some of man's most extreme individualistic or selfish traits.

Before the advent of sex, food was the only urgent need

of life. Now more is required: sex pleasure, sex selection, finer foods, the sense of beauty, personal adornment, the urge for ever more and more expression. Sex has produced *ideals*. There are different characteristics for male and female. Sex has lifted man above the commonplace, but it has also been the greatest source of brutality. "Man has always been most brutal to himself in the name of the ideal." Castration was one of the first surgical operations and most often done in the name of religion. In early ages children were castrated and thus prepared for the profession of eunuchs or slaves. In all ages it has been a religious rite by some fanatical sects. Even today there are cults in Roumania and Russia that practice castration. To a scientific mind it is very difficult to conceive of a Creator that would love the handiwork of His consciousness any the more for its being mutilated. Yet in all ages there have been fanatics that believe in perverting the natural expressions of the Great Creative Force.

Castration of boys before puberty retards ossification of the long bones with consequent enlargement of the stature. The lower limbs become disproportionately long. There is also adiposity. The larynx is not so prominent, and the voice remains high-pitched. Hair fails to grow on the face. The external sex organ remains infantile, and there is little or no sex feeling. Mental sluggishness prevails and the eunuch is lazy, suspicious and undependable.

Removal of the ovaries is followed by corresponding effects. If performed before puberty the characteristic feminine attributes do not appear, the girl tends to become mannish in type, the accessory organs of sex fail to develop fully and menstruation does not occur. In women, after the age of puberty, removal of the ovaries is followed by changes characteristic of the menopause.

The castrated male becomes female and the castrated female

becomes male in type. Experiments on these beings have shown that if an ovary is implanted in a male eunuch his general characteristics become female to a great degree. If a testicle is implanted in a female eunuch the characteristics will soon become male. If an ovary is implanted in a female eunuch the person will take on the whole appearance and characteristics of the female. If a testicle is implanted in a male eunuch the functioning of the male will be brought into expression. Eunuchs have more brittle and weaker bones than normal individuals.

The normal man is one with normal male gonads. The normal female is one with normal gonads of the female. The manly man and the womanly woman are the normal functioning man and woman.

The ovaries regulate the lime distribution in the female. Excessive pregnancies cause the terrible cases of osteomalacia or soft bone deformities that are so common in the densely populated districts of Europe and Asia. The frequent pregnancies use all the lime reserve and the bones suffer. Many women suffer with tooth trouble during pregnancy. In the male the testes (also called interstitial glands) regulate the lime of the bones and their strength and stability. The powerfully boned male is usually very virile sexually.

Some of the endocrines act as accelerators to the sex glands and some act as inhibitors. The thymus is said to hold the sex in abeyance and the adrenals accelerate the sex expression. The thyroid and pituitary also play an important role in the expression of sex.

The prostate gland, which is the storehouse for the seminal fluid, lies at the base of the bladder and surrounds its neck. Its complete function is not understood, but it must have some influence upon the nervous system for when it becomes inflamed the man becomes irritable, despondent and even suicidal. The author has restored many men to normal activity and function by treating the prostate. Sexual excesses are supposed to be largely the cause of enlargement of the prostate, as also is gonorrhea with its aftereffects.

Prostatic hypertrophy which tends to appear in the later years of life is attributed directly to hypersecretion of the male hormone. This hypersecretion is due to overstimulation of the testes by the gonadotropic hormone of the pituitary. Sexual indulgence or sexual excess has probably nothing to do with an enlarged prostate.

The ovaries are supposed to erupt an ovum every 28 days which is taken up by one of the Fallopian tubes and conducted to the uterus where it must meet the male germ (spermatozoon) if a new life is to be started. There is no stronger urge expressed in life than the effort of the male and female germs to meet. The breasts play an important part in the female expression. They form the food for the newborn child and they have an endocrine faculty that aids or normalizes the menstrual function.

There are periods of sexual desire and activity which in animals occur once or oftener in each breeding season. The first cycle commences at puberty. In women they are represented by the menstrual periods. During these oestrual cycles the ovum or egg is maturing and getting ready to be discharged for fertilization or impregnation. At such time, certain functional and organic changes take place in the internal and external sexual organs, including the mammary glands (breasts).

The uterus is the female sex organ where the foetus (child) is developed and prepared for its advent upon this sphere. Just what the endocrine influence is, science has not yet found positively, but we do know that when a woman has an inflamed uterus, she is irritable and usually depressed and generally neurasthenic. There is an intimate alliance between the postpituitary and the uterus. A few drops of postpituitary extract injected in the circulation will cause intense contraction of the uterus. This knowledge has been of great value to the obstetrician in the delivery of the child and holding the tone of the viscera. The reasoning mind cannot yet conceive how or why at just the correct moment there is freed in the system an excess supply of this post-pituitary endocrine substance that finds its way to the uterine cells and causes rhythmic contractions which expel the child and contracts all those blood vessels of the uterus that have been doing such heavy work for nine months.

The Thymus Gland

The thymus is a ductless, or a ductlesslike, gland situated below the thyroid in the upper anterior mediastinal cavity (behind the upper chest bones). It usually consists of two longitudinal lobes, joined across a median plane. Each lobe is made up of smaller divisions called *lobules*. Each lobule consists of an outer portion or cortex and a central portion or medulla.

The thymus of the infant is relatively large but during later childhood its weight, in relation to body weight, gradually decreases. It weighs from 25 to 40 grams. At about 11 to 14 years of age, there begins a regression or involutionary process within the gland. The regression is very slow and continues throughout life.

There is not enough known, as yet, of the action of the thymus but it seems to be the dominating gland of child growth before the time of puberty. It inhibits the activities of the testicles and ovaries. Castration causes persistent growth of the thymus. Removal of the thymus or its inhibition by the X ray hastens the development of the gonads. The continuance of the thymus after puberty causes peculiar actions of sex expressions. Repulsive and degenerative practices come invariably from thymo-centric persons. The thymus prevents differentiation and stops the transforming into positive sex expressions, either of male or female. Feeding tadpoles thymus substance prevents the evolvement or differentiation of the tadpole into a male or female frog.

In thymo-centric persons we get the homosexual cases. The male does not become fully male and as there is so much of him still potentially female he will care more for the society of the male than for the female. The female will still be potentially male and so enjoy the society of the female most. Our degenerates and criminals come mostly from thymo-centric persons. The thymus seems to be the child bodybuilder, supplying many of the elements that build the structure. It begins to regress at puberty so it is supposed that the gland is the gland of childhood growth. In animals whose thymus has been removed the lime or calcification processes become retarded. The thymus seems to dominate the lymphatic system.

The Parathyroids

The parathyroids are composed of four tiny glands, as large as wheat seeds, in or near the thyroid gland. Removal of the parathyroids is followed by great excitability of the nervous system. The reaction is similar to that of an overdose of strychnine. They are called the glands of tetanus. The chief function seems to be to control the calcium metabolism, or the lime salts of the system. These glands seem to be necessary to the steadiness of nerve and muscle control. They seem to be the agents of detoxication. There is always a lack of parathyroid endocrine secretion in lockjaw, epilepsy, paralysis agitans and convulsions of epilepsy.

Other Glands

The pancreas is the controller of sugar metabolism. This gland has a duct that carries its secretion to the intestines where its enzymes control and complete digestion.

The spleen is a ductless gland of which little is known. Known functions of the spleen are: (a) the final destruction of blood cells; (b) storage of blood; (c) manufacture of lymphocytes.

It is believed that only the fragmented dead, effete, and senile red cells are disposed of in the spleen, which acts as a graveyard for the dead and dying red cells.

Most red cells are destroyed within the arteries and veins before they reach the spleen. About 10 million red cells are destroyed in the blood vessels every second of the day and night. This loss must be replaced every second. Blood cells are made in the bone marrow and the life of a blood cell in man ranges from 25 to 100 days.

The spleen stores blood for emergency purposes and for special needs. This blood has a much higher red cell content than that of the circulating blood. The spleen gives up some of this blood to the general circulation when the individual exercises; it would also do so in all cases of hemorrhage or in any case of poisoning.

When cats are excited, the blood in their vessels is increased 25 per cent by the spleen's discharging its blood into the general circulation. Certain drugs, by contracting the spleen, will cause a discharge of this stored blood into circulation. Also, if one goes from a cold or cool temperature to a hot climate, some blood discharge will occur. This will be returned to the spleen when the individual returns to the cool or cold climate.

The third function of the spleen is the manufacturing of lymphocytes or white blood cells. These are the watch dogs, or the guardians, within the body to destroy bacteria and other foreign matter that may gain access to the blood stream or solid tissues.

The liver is one of the most important glands of the system. It is the storehouse and the clearinghouse. It is here that the food is finally prepared for its advent in the circulatory system and where the broken-down particles of the system are renovated and again made fit for use if possible. It is a gland of internal and external secretion. The external secretion is bile and is thrown by a duct into the digestive tract. The internal secretion is sent directly into the circulation.

Following are some of the many specific functions of the liver: (a) the production of bile which aids in the digestion and absorption of fats and also prevents putrefaction in the bowels; (b) the completion of the digestion and metabolism of food; (c) the detoxication of poisons; (d) the production of fibrinogen and prothrombin, which are essential to cause the blood to clot in case of hemorrhage; (e) the production of heparin which helps to keep the blood from clotting in the circulatory system; (f) the storing of food—namely, fats, proteins, and carbohydrates; minerals, such as iron and copper; and vitamins such as A, B, D; the liver manufactures Vitamin A from carotene; (g) the regulation of the blood volume; (h) the production of much of the body's heat.

The kidneys are glands of secretion and excretion. They secrete uric acid, ammonia, and hippuric acid. They excrete the products of secretion as well as others, including sugar, chlorides, urea, creatine, creatinine, water, potassium, calcium, sulphur, magnesium, phosphorus, fatty acids, pigments, and some additional waste products.

Most of the detoxication is done in the liver, but a lesser amount is carried out in the kidneys. It is very doubtful if the kidneys have any endocrine function.

Man can live in a state of well-being with but one healthy kidney. If the second or surviving one becomes much impaired, sickness and death soon follow.

The lymphatic system commences as a meshwork of very minute vessels, capillarylike, which drain the tissue spaces of the body of broken down particles of tissues. By the joining of small lymphatic vessels larger ones are formed. These in turn, by receiving tributaries along their course, gradually increase in size and finally form the right lymphatic and thoracic ducts.

These empty their lymph into the blood stream by way of the right and left subclavian veins respectively.

The lymphatic vessels in the intestine are known as lacteals. The lymph nodes or glands are important structures for the defence of the blood against the invasion of bacteria or other harmful agents travelling by the lymph channels. When an infection of a part—a finger or toe, for instance—lying distal to the gland occurs, the gland becomes inflamed as a result of the localization therein of some of the bacteria or their toxins carried in the lymph. The gland is filled with white blood cells called *phagocytes* which attack and destroy the invading organisms. In this way a barrier is raised against the passage of harmful agents, bacteria or toxins, into the blood stream.

Too much food and refuse, with poisons of some kind, often cause enlargement of these glands and make good foci for tuberculosis and cancers. The writer has seen many cases that seemed like tuberculosis of the lymph glands entirely recover from regulating the food supply and stopping autointoxication.

The salivary glands are the glands of the mouth that pour out the saliva and start the starch digestion of the food and maintain moisture of the mouth and throat.

CHAPTER IV

An Introspect—The Mystical Laws

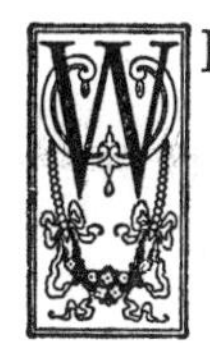
HEN one studies the action of the physical expressions of man it almost makes one feel that man is purely animal or physical or mechanistic, but more careful analysis proves that man is more than merely physical. To play, to work, to reproduce are common to man and animals. To *create* is human only and is the link between the human and the Divine. As we said before, we cannot conceive of a creation without a Creator. How could we have all the immutable laws that control all creation without a consciousness greater than our own to evolve all this? We have, as yet, sensed no human consciousness that seemed able to evolve all the vast universe. That it was all chance is equally impossible to conceive. We cannot help sensing that some vast force or consciousness is expressing itself.

Man seems to be a consciousness within this vast consciousness that is ever trying to express himself, sometimes intelligently and more often blindly. We know that there is an intelligence or consciousness within man that is higher than the mere animal. To find that higher consciousness and intelligence is man's supreme work at present. "Know thyself" has been the adage for centuries. Our great desire should be to know and when we *know,* to transmute the knowledge to higher planes of expression.

As we learn the laws of the lower forms of life we are learning the actions of the higher forms, for the teaching "As above so below and as below so above" becomes very vital as we gain more and more knowledge and wisdom. The deeper we peer into the mysteries of Nature the more we are reminded that the four fundamentals mentioned in the foreword are true and that every phase of life's expression is within the fundamental laws.

Man is linked by the ties of cell, blood, and bone to every expression of life in the sea, jungle, forest, plains and cities. Man is a branch of the tree of animal or beast nature. But there is more than beast there. Every cell of the body has its consciousness and knows what to take as food and what to eliminate. The liver cells have the liver-cell consciousness. The muscle cells have the muscle-cell consciousness. The skin cells have a consciousness quite different from the liver, muscles or any other cells and know how to act as skin cells and would be lost if transferred to the liver or the muscle. Man is a differentiation from other animal forms and consciousness. He is all right as a man but all wrong if he tries to be a lion, or a horse or a fish. Man must be true to the impulses or consciousnesses that have built him so far in his development. He is still bound by his needs and the needs of his ancestors which function through his endocrines—the ductless glands. His hates, loves, superstitions, food desires, and lusts still dominate him largely. Slowly, oh, so slowly, he is seeking freedom from his limitations. Reason and spiritual guidance take long to establish. That something called the Spirit of Christus is so slow in developing in man.

As we begin to understand the full evolving force of the endocrines we will gain true freedom and soul growth. Reason and the Divinity within must become the guiding forces of man. Man must have freedom for himself and for his fellow man. In the ages past man has enslaved his fellow man and lived a predatory life. There have been those high up and those far down. It is time for the law of normalcy and this will come as we gain knowledge of the laws of building normalcy in physique, mind and spirit. We must know the laws of involution, evolution, and continuous life. As we study the endocrines we know that man can be the architect of his own destiny. Mystical science will do more for the upbuilding of man than has been done by the religions of the world. Religion is the expression of but one phase of man's existence. Mystical science, such as the Rosicrucian principles, reaches all the phases of man's functioning and being.

Science has taught us positively that man's physical form and mental attainments are controlled by his endocrines which are the products of the ductless and other glands of the body. The length of limbs, the kind of face, the shape of the pelvis, the color of the skin, the tone of the muscles, the mental activity, the inheritance, all are effected by the endocrines. Races are small or large according to the actions of the endocrines which in turn are influenced by environment.

Napoleon shortened the stature of the men of France by the killing off in wars of the biggest men so only the smaller men could breed and reproduce. Also the nervous strain on the adrenals and other glands of both men and women inhibited the proper growth of the offspring.

How shall we build the bodies and characters of men? Do you know of a better way than to study and know the orderly arrangement of the forces expressed in man—which is mystical science—and being guided by the truths there found?

So far we have treated the glands as if each stood out separately from the rest, but *they never function separately*. Each influences the rest of the chain. A physical shock or mental shock will soon involve all of them.

The endocrines are the directors in the corporation of organs and tissues and consciousness of the being called *man*. There are subcommittees that control certain organs. The growth of the brain is presided over by the adrenal, thymus, thyroid and pituitary. They decide the size, the number of cells, convolutions and speed of its chemistry or action. The sugar metabolism is presided over by the pancreas, adrenals, liver, thyroid, and pituitary.

These glands or directorates may be cooperative or antagonistic. The thyroid and thymus are antagonistic. One inhibits the other (this is illustrated by feeding the tadpole thymus to prevent differentiation and the feeding of thyroid to speed his development).

The thyroid and the pancreas are antagonistic.

The thyroid and the ovaries are cooperative. The pituitary and the thymus are antagonistic. The pituitary and the adrenal cortex are cooperative on the brain and sex cells.

The ideal condition of mind and physique is for all the endocrines to have a correct balance. This is called *Harmonium* by the Rosicrucians, since the interplay is normal, and requires normal environment. While normal environment would seem like Utopia, yet we must have the creative mind to try to develop the correct environment. We have no other way of functioning or expressing self, but through the body and mind. Mind is but a consciousness. It is not something separate from our being.

We have two minds in our being: One is the reasoning mind and is called the conscious mind; the other is the subconscious mind. It functions when the reasoning mind is at rest and is the building force of the body and mind. This subconscious consciousness is the positive impelling force of man and is expressed in the endocrine system. It is called by scientists the vegetative system, and is supposed to be a lower form of action. The vegetative or endocrine system is the consciousness of the ages of involution and evolution of man. The impelling forces within man come from these endocrine expressions. Our emotions are the actions of the endocrines. The reasoning mind has no emotions. In fact it is a question if man's mind is as yet capable of any great reasoning, for all his reasoning is so tinged by the impulses from the endocrines that they overshadow the supposed reasoning. Most of man's reasoning is but seeking proofs to go on thinking as he has in the past. If he is a Christian he seeks proof for his belief and finds the answer in the impulses of the past recorded in his endocrines. If he is a Jew the process is the same, as also it is if he is a Buddhist or a Moslem. He loves and hates not by any reasoning power of the mind consciousness but by the deep impulses of the endocrines. We unconsciously absorb the arguments that come in our everyday contacts. That is environment.

We must make a new era—an era where we will know that we have evolved. We have been dominated by bodily processes, animal impulses, savage traditions, infantile impressions and numerous traditional and conventional reactions, and now we must use clear reason and clear thinking and transmute the knowledge of the past into wisdom and spiritual life for the future. We must leave behind the exhortations of the past as to the impurity and baseness and vileness of the physical body. We must learn that a clean soul must have a clean body to function through, if the functioning is to be clean. Men, like animals and plants, go on, generation after generation, living as their progenitors have lived, for the vegetative system has reasoned very slowly and has acted only from necessity, but we have arrived where we know that the past is not the sole standard for the future. Our past is too full of fears and hates. We need to become truly mystical and seek more light.

We said before that the brain is not the exclusive seat of the mind. It is only one unit of the intelligence system of the body. The glands are the tuning keys that lighten up or tighten up the driving forces of the system. This tone or driving force of the system is called the kinetic drive and is registered in the consciousness of the glands which is the so-called subconscious mind. This system is interrelated by chemical processes and through ganglia or nerves of the sympathetic nervous system to the solar plexus and other plexuses and the brain. Your wish or your will is not a thing of the thinking mind, but is a matter of the standards of the glands or endocrines, or the so-called vegetative system. If we know a man's internal secretory composition, we can predict very accurately the physical, mental, and spiritual make-up of the man and also the general lines of his life. Diseases, tastes, idiosyncrasies and habits can be predicted.

The past action of a man will give his endocrine standing. Customs, morals, ethics are the endocrines. Our evolvement is the story of the ductless glands, and the Cosmic mind will be raised only as the power of the individual mind is raised through the functioning of the endocrines and the reaction upon the mind. Our thoughts affect the endocrines as also the endocrines affect the mind or brain. Foul thoughts affect the endocrines. Foul and decaying products of digestion act upon the ductless glands or endocrines and may plunge the person into deep melancholia. Narcotics may for a time transport the person to realms of bliss through exhilarating action upon the conscious and subconscious forces. A sudden word or shock may act as poison. Loves, hates, fears have their depressing effect or exhilaration as the case may be. Diseased endocrine glands will affect the thinking mind of man and color his thoughts. Diseased gonads will cause the mind to dwell on matters sexual, and will cause irritability or a state of fear. Diseased thyroid will cause depression if hypothyroid, and a greatly excited condition if hyperthyroid.

▽ ▽ ▽

CHAPTER V

Types of Endocrine Persons

THE normally balanced endocrine human being is rather rare: There are as many different types of unbalance as there are endocrine glands. Our environment and possibilities for expression are still too inhibitive. When man finds the freedom he is ever seeking, then will he find balance.

The Thyroid-Centered Personality

The normal thyroid personality has bright eyes, good clean teeth, symmetrical features, moist flushed skin and a temperamental attitude toward life.

The hypothyroid or lack of thyroid personality is usually below the average in height with a tendency toward obesity. The complexion is sallow and the hair dry and the teeth irregular. The extremities are cold and bluish, the circulation poor.

The intellect is pretty sure to be dull unless the pituitary is large; the mind may be fine but the energy will be always lacking.

The hyperthyroid (excessive thyroid) person is the ruddy, live-wire type—active and energetic, fair complexioned and magnetic. The thyroid regulates the speed of living. It promotes the activity of the adrenals and in that way produces the activity of the system. The thyroid-centered person is usually the restless, active, perpetual worker who gets up early, works all day until late at night, goes to bed and plans the work for the next day, and then complains of insomnia. These persons are very susceptible to shock and worry or grief and their friends will be surprised that such energetic persons should so quickly become invalids and subject to various forms of psychosis, especially melancholia. Shock inhibits the endocrine secretions, iodine, phosphorus, and arsenic of the thyroid, and with it goes the break of normal interplay between all the other endocrines, and an especially live wire becomes derelict on the ocean of emotions.

Pituitary-Centered Type

The hyperpituitary or the sufficient or dominant type is usually large, with large, long bones, and the frame is dominantly "bony." His eyes are wide apart, face broad, teeth broad, large, and unspaced. The chin is usually square and protruding. Feet and hands are large and there is early growth of hair on the body, and a thick skin. This type is usually well sexed and aggressive, precocious and self-contained. There is usually an acute sense of rhythm. The features are not usually symmetrical. Abraham Lincoln is the extreme type of the pituitary-centered person.

The hypo or inferior pituitary type is small, sometimes with very delicate skeleton, is rather prone to fatty tissues and weak muscles, has prominent or protruding upper jaw, dry skin, small hands and feet, abnormal desire for sweets, subnormal temperature, pulse, and blood pressure, with poor control of the vegetative (ductless gland) system; he is mentally sluggish, dull, apathetic and backward, losing control quickly, crying easily, discouraged quickly and having no stamina.

In the pituitary-centered person much depends upon the sella turcica or the cradle of the gland, whether large and roomy or small and restrictive. The degree of development of the other glands also aids or retards the pituitary.

Much depends on whether the anterior pituitary or the

posterior pituitary is dominant. In the male the anterior should be dominant, and in the female the posterior. Also with the anterior pituitary the testes should be allied and with the posterior pituitary the ovaries should be conjunctive or allied. The anterior pituitary and the ovaries dominant would not be a good combination. It would make the masculine woman. The posterior pituitary and the testes dominant would make the feminine man.

When the posterior pituitary dominates in a woman and there is a good ovary support, the build will usually be rather slight and delicate, the skin soft, moist, roseate. There will be a fondness for children with rather an emotional tendency, in fact the ideal feminine type of woman. The unstable postpituitary female is unstable in all of her expressions, craves excitement, constant change and a new pleasure every minute.

Wars, excessive excitements, excessive sex sensualities, improper dress, improper foods have produced many unstable postpituitary types.

Many men are postpituitary centered and are often the poetic type, the musicians and very emotional. They are usually short, round and stout. Here we have the henpecked husbands and lovers. They often are beautiful characters, but lack aggressiveness. They should be understood, not bullied. Many women are anterior pituitary centered and so become the aggressive type and fill men's places in the business world. They too should be understood and not abused.

Adrenal-Centered Type

Hairy, dark, masculine, primitive and strong. Here we have the slave driver for he has such a sufficiency of driving force he can drive others. Among the high salaried persons and men of great energetic positions we have the adrenal-centered type. The adrenals, in conjunction with the pineal, control the pigmentation or the darkening of the skin. The dark-skinned and red-haired persons are typically the fiery adrenal-centered persons. In persons who lack adrenal secretion, influenza and diphtheria are most easily contracted. They have a lack of immunizing force. The person who has a good adrenal supply with good thyroid and pituitary action can lead the world. He can be a master among men. Brain fag is often due to adrenal fag. The adrenal type among women is masculinoid.

Moles on the skin are the product of the adrenals. The adrenal insufficient is weak, irritable, lazy and likely to be neurasthenic, and has loss of appetite and a general lack of response to stimuli of all kinds. Growth is slow in an adrenal insufficient and he cannot be driven or hurried. Children who lack the adrenal secretion before puberty often awaken to good energy when the rest of the endocrines develop—especially the sex glands. This fact should be understood by the educators of the land. The children who have not a sufficient supply of adrenal secretion cannot learn well and cannot be driven to learn. They will lack also the iodine and phosphorus supply of the thyroid endocrine secretion and thus the brain conductivity or the registering force in the brain will not be there and you cannot expect a child to register impulses unless the medium for impression is supplied. You should no more expect a mind to register without the material on which to register than you would expect the phonograph receiving horn [recorder microphone] to register the vibrations of sound without the wax or composition plate for the needle to act upon. Educators have much to learn from the endocrine system. Teachers have acted upon the supposition that the brain was a recording disk upon which they could pour impulses and have them register. The iron, the phosphorus, the lime salts, and all the subtle agencies of the endocrines must supply sufficient balance or the brain and general kinetic forces cannot act. Schools, like churches, have been places for inhibitions instead of activities along the natural paths of expression. The hope for the future in education and religious expression is for the doctor, the teacher, and the preacher to pool their knowledge and forces and educate along the lines of least resistance, which is nature's way toward full expression of consciousness.

Gonad-Centered Type

This subject cannot be adequately handled in a book for the general public until the general mind is less sensual in its outlook.

It is the sex glands that make the male and the female types of persons. Masculinity and femininity are expressions of the interplay of all the internal secretions. The testes and ovaries only give certain tendencies. There is no absolute masculine or absolute feminine, for there is still much of both within each of us. This will be understood if the reader will remember the action of the pituitary gland in its anterior and posterior aspects. The testes and the anterior pituitary make the dominant masculine person and the posterior pituitary and the ovaries make the dominant female. When the thymus gland has prevented the full development of the differentiated sex glands, we then get continuation of the two sex expressions within the one person. Here we have the homosexual person. (See thymus gland.)

Sex seems to be due to chemical reaction, and depends upon the number of chromosomes in the cell of the egg. The male has 22 chromosomes and the female has 24 chromosomes. Also lime salts play a great or dominant role in the development of man and woman. The male is more stable in lime salts action. The female is more unstable because of the periodicity of her life—menstruation, pregnancy and lactation. These draw on the lime salts reserve. The male is bigger and stronger because of these facts. The eunuch is one in whom sex has never developed because of castration, and he is always childish in feature and mind yet reaches senility when yet young. He does not have the lime salts which make for stability. There are also infantiloids, who are persons that have not been castrated yet are infantile in sex development. They are much like the eunuch. We also find that infantiloid tendency is invariably toward homosexuality. Homosexuality is defined as the desire to associate with one's own sex or where sex pleasure is most greatly expressed with one of the same sex.

History tells us of various cults of homosexuality. In many cases homosexuals were produced by unnatural means of inhibitions and sex irritations until all natural sex expressions were impossible. These were used in religious orders or as prostitutes for orders or sects. Even to this day there are religious orders that consider it a part of their religion absolutely to inhibit every sex impulse, and while these orders contain many brilliant men, for they work off their sex energy in mental attainments, the average person of the cults is nervous, and suspicious of all men and women. He becomes very self-centered and very rarely a great leader of men. Real "*he*" men do not readily follow a feminine or suppressed type of man. Neither will really feminine women follow or respect the feminine type of man. Men with strong anterior pituitary, strong adrenal glands and weak sexual glands are apt to be very cruel and destructive in their expressions. Repression in the male will often start the feminine trend and in females start the masculine trend.

The ideal normal man is the man with strong sex power and a well-developed anterior pituitary with the balancing power of the postpituitary normal, and strong adrenals. This is the man who will be creative in his work, kind in his actions and yet have a driving force that will meet and break down all barriers to his progress. This is the type of man that will be loved by both men and women. He will be a man's man, and also a woman's man. He is the type that will live up to higher planes, for he himself will be constantly striving for that which is best and highest. In him will be the poetic genius that drives him ever on and on. He will be able to realize the four great principles of life and can transmute his knowledge into building his physical, emotional, mental and spiritual expressions to the very highest. He will become the superman and then the master.

And similarly will evolve the woman who is truly feminine and well balanced in the sex and other endocrine glands. This truly feminine woman will be the medium for unborn lives to enter this sphere and through her, in conjunction with the perfect type of male, the incoming life will not be handicapped but will be free to evolve as the Creator has directed.

The history of man's expression on this earth has been one of cycles of profligate abandon to excesses followed by periods of terrific inhibitions. After an age of wanton excesses there went forth from *Mt. Sinai* the command of "Thou shalt not," and due to misinterpretation there followed a period of asceticism and inhibitions from which we are still suffering. These thunderings from *Mt. Sinai* have been interpreted to mean that all physical expressions were impure and unpleasant in the sight of God, and yet this same God by the same teachings caused all the impulses that drove men on to expression. Science will soon step in and by natural teaching will bring mankind to a realization that every expression of consciousness may and must be pure and uplifting. Man's creative force is essentially a pure force and just as necessary and sinless as the eating of food. Eating of food may be as unclean and excessive as the sex force. Young men and women have not been taught to know and live their creative lives. They have learned to sneak and steal and abuse their God-given forces. We have treated our hogs and cattle better than our children. To teach children simply to suppress all their lives is as productive of results as to try to blanket the crater of Vesuvius to keep it from erupting. The eruptions will come and the cities will be destroyed.

Science will teach how to stabilize man's excessive vegetative (endocrine) system and produce a race of balanced men and women. Our instincts, which are the subconscious intelligences of the ductless glands, will lead us on to normal expression if they are not wrongly inhibited. We have been taught to live "by faith alone" and I do not belittle faith when I also demand that I may be led to live by the law of all life, which is the law of action and reaction, or the law of cause and effect. Rosicrucians call this "Karma." Man can understand the creative laws if he will only study and learn.

The secondary characteristics of the male are: Hair on the face, skin coarse and lean, muscles powerful, bones heavy, bass voice and generally aggressive.

The secondary characteristics of the female are: Hairless face, skin fine and plump, relatively weak, bones light, treble voice and usually reserved. Woman's expressions are rhythmical. They act with the tide and moon phases. This may be traced to the posterior pituitary which in turn may be traced to the time when the pineal gland was an eye.

That the creative force of man is largely centered in the sex endocrines is proved, for in all eunuchs that were so made before the sex ever developed, there has been no creative energy ever developed. In the past there have been many cults to inhibit sex expression but for the future, science will teach that he who inhibits all the endocrine impulses of his being will be as impure and unholy as the being who indulges in excesses. The mind and higher spirit of man cannot develop when the prime creative force is destroyed. (In a later chapter we will show that sex expression is not mere copulation.) All the endocrine expressions should be pure and not excessive. Man cannot evolve through lusts or excesses nor through inhibitions.

Just at present there is a frantic effort being made by some men and some doctors to find the perpetual fountain of youth, and some think they have found it in the gland transplanting process. It is the usual foolish effort of men to outwit the laws of nature. Transplanting of an ovary or a testicle may give added impetus to the sex expression, but in the long run it must only wear out the general system the faster. It will be the plaything of the idle rich and sensational doctors for a time, and it may possibly add a few months or even a year or so to the sense pleasures of the few who can pay for the mutilation of some poor unfortunate, but the immutable law of "Karma" will not be made sport of, nor can it be cajoled by money. The "Temple of Myself" is holy ground to our way of thinking and the effort to find ways to continue mere sensuality is very disgusting to us. The endocrines that are not abused will function well until very old age.

We do not wish to say much about treatment with endocrine products, which is getting to be the rage at present by doctors and by quacks or proprietary concerns. We do wish to give the warning that it is best to go very carefully. If you have any endocrine trouble consult a doctor who really has made a study of the subject. The mere feeding of a gland product cannot correct the trouble. There is some law of life being violated when the supply is not normal. The thing to do is to find what law is being violated; correct that and then find the best method of arousing the forces within the glands. It may be that all that is needed is mental or emotional calm. Perhaps exercise or more sunlight and air is necessary, or it is just possible that you are using your vitality too fast and are in the state of "Fight or Flight." It could be that you do need some endocrine products, but it is not safe to use them indiscriminately. The science of supplying deficient endocrine substances is not fully developed. The object of this story of the endocrines is to make the laity a little more familiar with the building forces within the human temple and the sacredness and the possibilities of life's evolvement.

Thymus-Centered Type

Up to the time when the permanent teeth are cut, the thymus is the dominant gland. This age is six or seven years and here the child form is very much alike in both sexes. Then slight differentiation begins, though no marked changes take place until the time of puberty. At this time the thymus functions less and less and the sex and other glands begin a greater development. When the gonads are fully established, the thymus is supposed to have become inactive. But often the thymus goes on functioning for some unknown reason (probably some inherited trait or lack of sex growth), and then we have a person whose whole life will be dominated by the thymus gland (thymo-centric). The features will stay rounded and childlike. The ruggedness of the sexual or pituitary type will be missing. In this type we get the "angel children" that are so delicate and fair of skin and features that they seem to be not of this earth, and their movements are all grace. Novelists seem to delight in describing this type of child. It is not a normal child.

The thymo-centric is handicapped for life's stress as the body is usually not strong and is subject to being easily shocked.

It is the proverbial "good" child that "dies young." They often die suddenly and without apparent reason. They do not stand operations well. In one of this type, puberty is delayed and is difficult to establish.

There is still some disagreement as to the action or lack

of action of the thymus after puberty. The most common opinion seems to be that a persistent thymus after puberty tends toward producing the feminine expression in the male and the masculine expression in the female. That is, a partial castration takes place. Increasing the interstitial secretion of the male thymo-centric will establish a better masculinoid, and supplying the female interstitial secretion will establish a better balanced female. The thymo-centric is to be pitied and science will soon come to his aid. The thymo-centric will often wonder, as do all his associates, why he is not like others. There is the peculiar complex that the male thymo-centric will want the society of the male more than the society of the female which is not the case with the normal male. The normal male naturally seeks the female companion. The thymo-centric female will have the complex of preferring the female society to that of the male. The normal, fully developed female will naturally desire that the male will seek her, and wishes male companionship.

Homosexuality (desire for one's own sex) may be concealed, but often is frankly conceded. This sex complex complicates the social adjustment of the person. It often and usually makes it difficult to train the boy in the male expression of his life, be it play or work, and the girl stays hopelessly "Tom boy." The pituitary also seems to be unable to function properly to assist the person to correct reactions. This child will be apt to be a late childhood bed-wetter and will have a very small sense of the proprieties of life. In this class we get the pathological or unconscious liar—the child that will steal, promise not to do it again, and in a few moments do the very thing again. There is no sense of responsibility. He may not be vicious but is just generally irresponsible. Even the tissues are unstable and subject to tuberculosis, meningitis and all children's diseases.

The author personally knew a lawyer who was a typical

thymo-centric case. Even in common conversation the man would lie when the truth would have been far better. He had the typical rounded child features but was of rather tall stature. He was well developed mentally and a good talker, but irresponsible. He was quite the Oscar Wilde type. His irresponsible habits caused him to be sent to the penitentiary. We now know that the man was not really criminally inclined, but this undeveloped side of his life was his downfall. The time will come when we will treat as sick, the morons, the endocrine deficients and most of the men that we now condemn to prison for life and make into brutes. The spiritual vision will give us the insight to the actions and reactions of life expressions.

We will some day know how to develop the sex and pituitary glands and endocrines so as to cure the thymus dominant cases. It is in these thymo-centric cases that we get men who love men, and women who "marry" women. The thymo-centric is apt to be generally weak and knock-kneed, flat-footed, fragile, with poor circulation and handicapped for life. It is claimed that alcoholics, drug habitués, criminals and degenerates belong largely to this class or type. There is no stability, and so an ever-restless seeking for *something* is fruitless, and satisfaction of desires is never accomplished. These are the misfits of life. They do not fit into the normal scheme of things.

If the pituitary and the thyroid become well developed, the thymo-centric person may become rather brilliant. The lawyer we just mentioned was very brilliant but very eccentric. We find a great many epileptics among this type. Napoleon was somewhat the thymo-centric type though his anterior pituitary gland was his driving force. When that failed him he failed. Napoleon was an epileptic. He had small sex development, and no real love for women. They were merely a convenience. His posterior pituitary was not well developed and his adrenals were, so he was cruel in make-up and brooked no sentimental interference. He was not religious. Many of the great adventurers and restless experimenters of the world were thymo-centrics. Mohammed was an epileptic.

It is claimed that our murderers and suicides come from this type of beings.

Oscar Wilde was another type of the thymus-centered person. He was brilliant and wrote some of the most beautiful things in the English language, yet he was sent to prison for his homosexual practices among boys.

Most thymo-centrics are not brilliant. They are more commonly the ordinary brutal type of misfits of life.

Improper mating of parents often is the cause of the thymo-centrics. As two positives will usually produce a negative, so if two pituitary-centered persons mate and reproduce, the offspring is very likely to be thymus centered. This may explain why so often the offspring of two very brilliant people will be mediocre and irresponsible. Or two money-mad adrenal-centered persons will have a puny, weak (mentally and morally), degenerate child. The normal offspring can only come from a mating of the normal male and normal female.

▽ ▽ ▽

CHAPTER VI

Continuation of Types

AS WE study the actions of the endocrine system we realize that it is due to the imbalance or abnormal action of these glands that we have the various abnormal human beings. When the building and activating forces of the ductless glands (endocrines) are normal we have normal human beings. We will note some of the outstanding peculiarities of unbalanced endocrines.

Pituitary-centered persons are liable to headaches and eye trouble, for in mental activity the blood is sent to the brain more intensely and the pressure on the pituitary is the result.

Nietzsche was very brilliant but unstable, pituitary-centered, and was subject to intense headaches. Later in life he was mentally unbalanced.

Darwin was a neurasthenic pituitary-centered person. His adrenals were very much lacking in power. After his gonads became less active, his adrenals became more energetic and he was able to do more work. Men and women often are more vigorous after the menopause.

Many neurasthenics are pituitary-centered with failure of normal balance of the thyroid, adrenals and gonads.

One of the finest examples of pituitary-centered persons of history was Abraham Lincoln, but he was endowed with wonderful balance in all the other glands. He was strong and powerful, yet gentle, tender, patient and kind. It is the unbalanced person that is erratic, cruel, coarse and undependable.

The time is not far distant when criminals, degenerates,

and the outcasts of society will be understood and taught to function in harmony with their surroundings or be reclaimed and fitted to take their place in society with normal men and women. A man or woman who is suffering from lack or excess of the endocrines, which cause terriffic urges, is sick and needs an understanding doctor, not a jailer. They spread disease as does a typhoid or a diphtheria carrier. Would you put a typhoid carrier ir prison? No! You would hospitalize him to cure the trouble.

So with people who have unnatural urges and tendencies. The cause of the imbalance will be found in the endocrine system. The cure will also be found there, and not in prison. Life is action and will be expressed. To find normal expression is man's duty and pleasure. Abnormal expressions of life do not tend to pure happiness. Normal expressions do. Every step of the daily routine of life, every phase of happiness, of thought or feeling, is an episode in the endocrine reaction of the individual. How can the mind work normally when the factors that build the mind are defective? The endocrines build the mind. The endocrines build the physique. Your evolvement depends upon the activities of the endocrines. You are building for eternity now. If physical science is true in its claim that nothing is ever destroyed, that it only changes form, and if metaphysical science is true in its postulation that life is continuous, then we must go on, reasoning from the law of action and reaction, or cause and effect, that the causes we start now will be the effects of the future. Who can tell us positively where the effect ends?

Our teachers and preachers, as well as our doctors, need to know the laws of the endocrines. The teacher who is able to understand the cause of the impulses or lack of impulses of the children in the schoolroom can lead and control the children through a grind that is not unlike the old cider mill where large, small, wormy, bitter, and sweet apples were ground to a homogeneous mass and squeezed until dry. We need a system that will grade the pupils much as a prune or orange grader, where every grade will be put where it belongs and can be properly handled. Under the correct system the culls can be taken care of and put to the use for which they are fitted. The analogy is not absolutely correct, for children are not like apples or oranges; they have personalities that need developing and must have a chance. If the thymus gland is at fault it can be helped by shrinking with X ray or by feeding it gland substance. If there is sex irritation the cause should be found and removed. If the cradle of the pituitary gland is too small (the X ray will reveal it), then feed tissue salts or better foods for developing endocrines. Study the thyroid for excess production or lessened supply of iodine or phosphorus or arsenic.

Psychoanalysts claim that all our urges come from the sex complex. There is still a doubt if this is true. But it has been amply proved that it is the sex forces that produce the creative and imaginative qualities of mankind, as well as his decisions. A castrated person has no will power or energy. He also has no discrimination, so we know that the gonads play an important part in will and judgment.

The constructive imagination is due to a good balance between the anterior and posterior pituitary gland, with correct sex balance. We have thyroid moods, adrenal moods, antepituitary, postpituitary, and gonad moods. When we achieve absolute balance of the glands we will no longer have these special moods.

There must be forces as much more subtle than the endocrine as the endocrines are more subtle than the mind and physique. May we not hope to find some day the intangible influences that capture the thoughts impelling us onward, and then learn from them how to build ourselves with positiveness? We are still too unstable.

Man has *evolved* through the unconscious, but it is now time to aid by using the conscious. A wish is never born in the brain alone for the brain has no power to charge itself with energy. It can only store and transmit, for the source of energy is in the endocrines. The ancient philosophies taught that devachan (heaven) was a place where the physical and emotional part of man had been cast aside and where man lived in the mental state alone. All desire was sublimated and life was just a contemplation. Thus man contemplated until the desire came to enter again the realm of expression for more knowledge and experience. Then he again took on an emotional or astral body, as well as a physical body, and descended to the earth plane where he mingled with the evolving people, gained wisdom, and learned to help his fellow men and to become a master among men. Thus he took numerous rounds on the "Wheel of necessity" and through vast cycles of experience he became as one who had all knowledge and knew God.

One cannot ruminate upon the workings of the human in all his phases and expressions without feeling that there are vast forces beyond those which science has demonstrated, and that we, as human beings, have a very intimate connection with the ALL EXPRESSION.

We believe that we have come closer to solving the connection between the here and there, through the knowledge of the endocrines, than through any other source. We get nearer to the soul of things. "Acuteness of perception, memory, logical thought, imagination, conception, emotional expression or inhibitions and entire content of consciousness are influenced by the internal secretions." (Berman) Soul consciousness lies just beyond.

Though no wish is born in the brain it is through the thinking mind that we will have to find balance and the final uplift of this plan of action. When the thinking mind is fully developed then we will be ready for meditation and the transmuting of the knowledge we have gained into wisdom. Through wisdom we may gain adeptship, which is real self-mastery. The brain or thinking organ is constructed and made active through the presence of iodine, which gives electric conductivity, and through phosphorus, which is one of the most vital ingredients of the brain. These are furnished by the thyroid gland; so it must be healthy. The creative energy is furnished by the sex glands and the pituitary gland, so it is important that these be healthy and normal in action.

Fear, anger, hate and love, courage and desire for service are of visceral and endocrine origin, but the brain or thinking mind is the recording place or the transmuting organ, and like the phonograph record it can only register the impulses that are sent to it and can only send out what it has meditated upon and recorded.

Hunger is not a cerebral manifestation. It is visceral. Fear and anger involve the adrenals. They are the glands of combat. Courage comes from a good anterior pituitary and strong adrenals. The maternal instinct comes mostly from the postpituitary as also do the social and some of the creative instincts. Sex libido and passions are related to the testes and ovaries. Sympathy and curiosity are functions of the pituitary. The instincts of self-display and self-effacement, of pride and shame, are of thyroid origin. Thyroid is an energy producer and we live fast or slow according to the state of the thyroid. Memory is due to a good iron content of the brain, and iron is a product of the thyroid. The pituitary seems, however, to be the preserver of memory. A child may have a good memory but poor judgment, for its pituitary and gonads are not yet fully developed.

▽ ▽ ▽

CHAPTER VII

Methods of Developing the Endocrine Glands

THE pituitary gland can be stimulated by deep nose breathing. The Rosicrucians have taught this for years and have proved it by their mystical exercises. The blood circulation of the nose and the base of the brain are intimately connected. Singing that vibrates the base of the nose and brain will vibrate and stimulate the pituitary gland. The ancients pronounced the sacred word so as to stimulate their vital forces. (The Rosicrucians still use definite "vowel sounds" for this purpose.) They even warned against pronouncing of the sacred word by one who did not understand the potency of its power.

The thyroid is the organ of emotion, and so calm and poise are essential to its development. Lately we have found that static electricity and the X ray can do much in stimulating and inhibiting the action of the thyroid. Well-balanced tissue foods, such as fruits and vegetables are necessary.

As the driving force of the adrenals derives its power from the thyroid iron, it is necessary that the thyroid be normal and the fear and anger element be controlled.

There are two ways of remaining young: One is by keeping the thymus and pineal glands dominant and thus remaining juvenile and undeveloped, and the other way is to keep the sex glands normal and able to function, and being fully mature. When society will attain a pure mental attitude toward life's functions and when the subject of normal health can be taught in the schools so the children will get a comprehensive view of the building forces within them—the endocrines —then will the oncoming races become more developed mentally and spiritually, for they will understand their impulses and will control and use intelligently the God-given forces within themselves. No progress can be made, however, until the masses will understand that the physical part of man and the physical functions can and must be as pure as any of his mental and spiritual expressions. A pure soul can function better through a pure body than through an impure or foul one.

The human body in its development unveils and reveals the records of your past lives. What does your body reveal to you? Purity, love, high-minded aspirations, sweetness of contact with your fellow beings, or lust, greed, hate, sensual gratification and ignorance?

If a child is well born and has the liberty to express itself, its endocrines will most likely be normal. Of course, the child must have normal mental and physical food. A judge said recently that it is physical energy that drives lads into mischief and crimes. Certainly it is physical energy that drives, to self-expression of all types, any child or grown person. The judge himself would not be a judge if he had not had a superior energy that helped him to claim the place he now occupies. See to it that the lads and lassies have a normal way of expressing their energies, for they are God-given and they too may be judges or occupy prominent places. Do not throw inhibitions about them until the forces within them drive them on to criminality. Boys and girls are wonderful imitators and they express life very much like their grown-up brothers and sisters. They are often not so discreet and thus get into trouble.

Glands can develop normally only under normal conditions. Energies pent up within a person will have expression either openly or secretly.

Next to the food impulse is the sex impulse. Sex is the creative force as we have said before. This creative force WILL NOT be denied if the person is to stay alive and active. To suppress it would make a race of eunuchoids. We want a race of active, virile men and women. How can we attain this? By giving the race ample opportunity to develop its creative energies. Sex is expressed not only in copulation. Sex expresses its energies in multitudinous ways—in play, fight, study, painting, singing, decorating the body, religious fervors and catalepsis, football games, baseball games, bridge parties, gambling, hard work, home life, etc.

Children and grown-ups who are denied all healthy forms of expression will seek it in secret, and that invariably leads to the wrong use of sex—masturbation and liaisons. These facts we know as true, for they have come from our having lived in several communities where the young were denied all dancing or any form of amusement, except for going to church twice on Sunday and sitting still for two hours of tirade. The only way the young could get together was by sneaking off when the old people were asleep, and the result was just as we have stated.

There will be mighty few young people who will not be eager to live splendidly, when they are taught the actual truths of life and of their driving forces that are trying to express themselves. They will give heed to the laws of cause and effect and will hold their sex activities to normal expressions, and thus gain control of themselves and their lives, and live beautifully. We must learn control and proper sex use. Ignorance and complete inhibition or complete abandon will not bring happiness. Knowledge and poise will.

All nature at this age is keyed to the law of sex. The flowers, the plants, the insects, the animals and the human animals are all living under this law. Why should we not study the law?

"The science of sex is to know how to produce the most perfect bodies. The philosophy of sex is to know the purpose of bodies and make the best use of them. The religion of sex is to lead quality to intelligently become unity." Science, philosophy, and true religion teach the purity of sex and this is our only hope of regenerating the race.

A scientist was once asked why he knew so much about the fly. He said it was because he put himself into the consciousness of the fly.

When we put ourselves into the full consciousness of man we will know more of man. We have been putting too much consciousness into gods, angels, fairies, genii, and supernatural states, and in lands, houses, food, fame, honor, clothing and GOLD. All these are good in their places but they are not the most vital things of our expression. We plead for a fuller consciousness of TRUTH, KNOWLEDGE, WISDOM and LOVE that will lead us on to greater unity with the INFINITE INTELLIGENCE. We must learn to correlate all the forces within ourselves. The endocrines are our building forces and through these we develop mental and spiritual or creative power, and all those who have knowledge along these lines will be able to build themselves into a finer state and help all the rest of the hungry humanity to evolve. We *must evolve*. It is the law. Why wait to be driven to evolvement, why not be aggressive in evolving?

We must learn the effects on all of us (as individuals and as races) of hate, fear, anger, jealousy, business worry, quarrels, shocks; and of hope, faith, happiness, laughter, service, interest, religion; and of alcoholics, tobacco, teas, and coffee, drugs that are not foods; and of all pure thoughts and evil thoughts and lusts and indiscretions; and of a well-poised life. We must learn to be honest with ourselves. You cannot fool the endocrines. You can radiate what you wish. The power is within you.

The Temple of your Soul should be a pure temple. It should be holy ground. It should be a temple of joy and happiness. YOU can make it the Cathedral of your soul.

We must get away from the destructive forces of the "Fight or Flight" evolvement. We cannot build the human system and consciousness unless we learn to stop inhibiting our natural forces. Sickness is but an inhibition of the natural forces of man. There will be no sickness when man finds mental and physical freedom. Wars brutalize men and stop the normal interplay of the endocrines. The endocrines will do their constructive work when the personality of man gets out of the way. To construct building forces we must use all the forces at hand. We need to study the food problem and the clothing and housing problems. If we are to utilize all possibilities to evolve supermen we need to know the qualities of the endocrine substances, of food, of medicines that are constructive, and of surgery such as is helpful; also of light and the electrical, and constructive and stimulating forces. We must learn to put aside the inhibiting forces of wrong thinking.

Many cults have within the last few years been built up on the need of freeing the mind forces. The medical man has fought these cults, not recognizing that the urge for the freeing from thought inhibitions was more deeply seated than any cult has yet expressed. Spiritual and mental healing have very often caused a calm within the person and in this great peace the endocrines did their normal work, and the person became healed. Functional as well as organic diseases have been cured by the calm of mental and emotional exaltation or peace. The ideal doctor is one who knows all the forces of the physical body and also the activities and power of the mental and spiritual and emotional planes. The perfect doctor or healer is like the perfect man—still in the evolving.

To recognize that the healing or building forces are within

man, and all they need is freedom for expression and maybe a little stimulation from outside forces, is a vast step in advance. The doctor or healer can only help clear the way for the forces within to work. The medical man has, by his teachings in sanitation and his scientific findings of the action of the physical and endocrine forces and psychological actions, given to mankind more than he can ever be given credit for. The mental healers have given a new impetus to truths that were lying dormant. All the findings of science and metaphysical science need to be weighed and used when good.

Disease may be inhibited by drugs and also by mental processes. But when the inhibition ceases, the disease is still there. Just now autosuggestion is the rage. In some cases it may do good; the pain may be inhibited, but if the laws that produced pain are not corrected, the pain will manifest again. A person can suggest to himself that he is on top of Mt. Hamilton and keep suggesting, but if he does not obey the laws that take him to the mountaintop he will never get there. He can suggest to himself that he will go to the top of Mt. Hamilton and thus start the forces that will take him there. Here common sense and reasoning force are needed.

We need to learn the art of loving, for love is the great constructive force, as the great teacher Christ so clearly taught. Most expression of so-called love is but sex sensuality. The art of loving, the art of being kind, the art of giving service, the art of *being,* need to be taught in the homes and schools and in the churches. The Rotary Club, the Lions' Club, and all the progressive businessmen's clubs need this teaching and realization as much as the art of booming business, selling the town for publicity, etc. Heaven could be started right here. All it needs is here.

Show us a man's mystical philosophy and we will show you how far that man has evolved. A man's mystical philosophy is his highest conception of life.

CHAPTER VIII

Examples of the Inhibitions and Exhilarations of the Glandular System

N THE foregoing chapter we mentioned some of the inhibiting forces such as hate, fear, etc., and love, service, etc., and some of the narcotics. This book does not pretend to be a classic or an exhaustive treatise. It only hopes to stimulate further research into the subject. To those of the laity who wish to pursue the subject further I would suggest Berman's *Glands Regulating Our Personality,* Macmillan Co. Also Lorand's books, and the writings of Soddy, Cannon, and Crile.

We will here show a few specific cases of the actions of the glands under abnormal conditions.

One of the most pernicious influences upon the glandular system is the excessive use of the cigarette, cigar or pipe. The normal action of breathing or inhaling air is to supply the iron of the blood with oxygen. Oxygen is absolutely necessary in all the digestive and reconstructive processes of the body. Scientists claim it is the life-giving force or the substance that IS, or produces life. All slow-burning fires or incomplete-combustion fires produce carbon monoxide gas, one of the most deadly gases known. One part in eight hundred parts of air will cause death to a person in one-half hour. The exhaust from the automobile sends out large quantities of this gas and in the early years of the use of the automobile many deaths were caused by starting the automobile in a closed garage. The slow combustion of the tobacco produces carbon monoxide gas and this is inhaled by the smoker. As this gas has a greater affinity for iron than has oxygen, the oxygen is pushed aside and the carbon monoxide enters the blood in combination with the iron.

This combination has a very deleterious action upon the glands of the system. This is why the boy or girl who smokes excessively rarely has energy. He or she is lazy as a rule, yet nervous and excitable and lacks real driving force. The special gland affected is the gonad. The oncoming race will positively be weakened by excessive smoking. Experimentation with animals has proved that animals subjected to smokes (simulating the smokes of the human animal as much as possible) will after a certain period of smokes, no longer breed. The time is coming when men will no longer sit in smoke-laden rooms and inhale and reinhale time after time smoke that has passed through other nostrils and mouths and think it is the correct thing. Any smoke that is inhaled whether cigar, pipe or cigarette has the same effect.

Alcoholics have a certain activity within the system because of the effect of alcohol on the glands. When a man takes a drink of alcohol he may feel exhilarated. Why? Because it is a poison and the one gland that is ever on the alert is the gland of taste and smell—the pituitary, the subconscious brain. It sends out an alarm or hurry call to all the rest of the glands to get to work to expel the invader of the sanctuary. What is the result when all the glands send out all their forces to strengthen the citadel against the foe? An exhilaration. The glands and cells of the body, that is, the subconscious cells, have a greater preserving sense for the man than the man's thinking consciousness has. If the man took only this one drink maybe there would be no serious results, but the ignorant fool takes another, for the first seemed to do him good, and another and greater call is sent out and so on; drink follows drink until the glands can no longer work fast enough. They become overwhelmed. The man is overcome by the poison and we say he is drunk. The glands are like faithful dogs and no matter how they are abused they will ever strive to save the system from serious harm. They even may get so used to the alcohol that they will refuse to work until prodded by a drink or two.

So with all narcotics and drugs. There are times when drugs and narcotics may bridge over a crisis, but curative medicine must have a food value or stimulation that urges the reconstructive forces of the body, which lie in the glandular system, to do their normal work.

Overfeeding and wrong feeding will produce a self-poisoning within the digestive tract (autointoxication) that will do about the same to the system as alcohol and narcotics. The beginning of Bright's Disease usually comes from an abused digestive tract that has to throw off so much poison through the kidneys that the kidneys become diseased and can no longer function.

This form of trouble is very common with the businessman of America. Remember that hate, fear, excitement or any strenuous depressing or fighting emotion stops the peristalsis of the digestive tract (the wave motion) and the food is not carried on and digested as it should be. There is fermentation and putrefaction, and the product of this fermentation (autointoxication) is carried through the system and all the glands and tissues suffer. Friends often write about an Inscrutable Providence that has carried a brother away. An "Outraged Providence" would be a better expression. When businessmen learn a little about the wonderful forces within them and maybe a little less about the fleeting dollar that cannot bring health or happiness unless some simple laws of Nature are observed, then we will have more life and still enough dollars. Money cannot bring happiness. It can only clear the way for happiness. Knowledge and wisdom only can bring real happiness. The peristalsis of the bowels and the constructive and regenerating forces within the liver are controlled by the adrenals primarily, and any shock or overstrain, physical or mental, especially mental or emotional, will stop the actions of the whole chain of forces.

Drinking, smoking and excesses of food are not a good combination to take to a business where a clear head and fine decisions are necessary. The example of the cat under the fluoroscope will illustrate the condition of the businessman's case. Feed a cat a barium meal and put it under the fluoroscope and you can watch the wave motions (peristalsis) of the bowels and as long as the cat is kept purring the peristalsis is not interrupted. But then pull the cat's tail or prick her with a pin and get her angry or distressed and the wave motion stops at once and is not resumed until the cat is quiet and happy again. It is the old "Fight or Flight" story. So with the man or woman who is happy and contented, the wave motion of the digestive tract will be normal, but as soon as there is grief, fear, anger, worry, or any of the inhibiting emotions which stop the glands (adrenals, thyroid, and pituitary) from acting, then the peristalsis stops and we have putrefaction, fermentation trouble, liver trouble, constipation, etc. As long as the businessman works under hurry, anger, hate, fear, jealousy, etc., he will not be at his best. He needs to learn to "purr." It is time for man in his evolvement to sacrifice the beast within himself and learn to utilize his inner possibilities. There are too many "Goliaths" (evil minds) clothed in brass armor (materialism) trying to conquer the world, and we need more "Davids" (pure minds) with five perfect stones (five perfect senses) in their girdles, ready to use in their slings, with which to destroy the "Goliaths." Man needs to slay the "Goliath" within himself and develop his five senses and still other higher senses.

The business world is very slowly learning the power and usefulness and beauty of cooperation. The slogan has been that "competition is the life of trade," but cooperation is the only salvation of trade. Suppose all the ductless glands vied with each other in competition, what would be the result? Death of the individual. The glands of the system are ever striving to act in harmony. It is man's ignorance and superstitions that keep him from working with this vital force that lies within the glands. Man must learn to act in harmony with the forces within himself. This is so trite it seems foolish to say it, yet the vast majority of people are absolutely ignorant of the vitalizing forces that allow them to express life—not only as individuals, but collectively as communities, states, countries, nations, etc. Man meets man in business. They still fight or flee. Man meets as nation against nation. The same fight and flight action. No cooperation. Always competition. Always in high emotion of fight, fear, worry, competition, trying to destroy the other one and getting destroyed. Animals seem to live in greater harmony than men and nations. Men, communities, nations, cannot evolve and grow finer when they are trying to destroy each other physically, and are destroying also the finer forces within themselves that would build finer communities and nations.

The ramifications of the actions upon the glands of the system that build us are so subtle that it is necessary that we know more about them. Every action of life of man is connected with the glandular or constructive force of the system. The fight between capital and labor is playing a very great score in man's evolvement. Capital is determined to control the activities of business. It gets hard and calloused in its attainments. Excessive money in families or groups tends to dissipation, sex excesses and a general dwindling of the vital (endocrine) forces of the family. These groups soon die out; that is, within a few generations. It also produces egomania, which is an exalted opinion of self and families, and autocracy. Here we get the old autocracy of kings and the nobility. The history of all these has been degeneracy.

On the other hand, we have just now the autocracy of labor to contend with. Another factor in labor and capital is the efficiency craze that turns out as much as possible. This has led to the attitude of the laborer taking no personal pride in his work. His work is simply slavery to him.

No man can be absolutely healthy, physically or mentally, if he has no joy in his work. Work without joy in the work brutalizes man. The pituitary, adrenal, and sex glands will become less active and general stagnation of the whole system will develop. Under these conditions man's lust will become aroused and, as the higher creative force of man is not aroused, the lower will dominate and *idealism* will cease within man. This may go on for some time, but eventually man must evolve and the awakening will come and woe to the forces that meet this aroused consciousness.

Some day we may get leaders of both capital and labor who will have vision enough to know that their interests are mutual and that work must be as joyous as the commanding of industries, and then will the endocrine forces of man again have the possibilities to evolve supermen.

Wars brutalize men and stop the play of endocrines. There is always a wave of crime and brutality after a war. The balancing forces of the endocrines have become disturbed. All Europe is in constant fear of war right now (1939). Wars never really decided any great matter. Wars are the result of man's endocrine unbalance—his lusts, his greeds, his superstitions, and hates. The emotional and mental state of Europe is in complete imbalance.

A religion that teaches fear inhibits the endocrines and the finer expression of the ideal. The teaching that man is born sinful tends to make him so. The temple of man is within his body. Here he expresses what is good and what is bad. This "Temple of Myself" is sacred. Within this temple the physical, emotional, mental, and spiritual parts of man express the glory of the Creator. We repeat, we cannot conceive this creation not having a Creator. Man is inherently pure and his endocrines will build purely if allowed to do so. We are here to learn to correlate the forces within us for they are Divine forces. We are also to learn to cooperate with the forces within our brother man and all our environment.

Scientists tell us that man uses only about one tenth of the cells of his brain. They are all there to use. Man is not more than one tenth civilized. The endocrines have built us so far but it will be the use of the brain by the mind that will complete man's evolvement into full consciousness. The endocrine glands will build us normally when we learn to use our minds so as to inhibit fear and all depressing emotions. Under the higher attributes of life the vibrations or the life expressions are normal and constructive.

Under a hopeful religion, a religion of peace and joy, the endocrine forces are upbuilding. In a happy and peaceful home the child will develop normally. In a home of quarreling and fighting, children cannot develop properly. Eating should be done under happy conditions for, as we have explained, peristalsis and correct digestion cannot go on under fighting or quarreling. In the average home of today the mealtime is the clearinghouse time for all the troubles of the family—the time when children are scolded and even punished. The writer once had a very sickly and nervous child to treat and could not find the reason for the condition. Happening to be in its home at mealtime for a few minutes, we saw the child eating nicely and it leaned over toward the mother and asked a simple child question and the mother, instead of answering the question nicely, said "Shut up" to the child and hit her over the head with a cup she had in her hand. We then knew where the trouble came from. We also found that the husband and wife were constantly quarreling and fighting and that was due to sex incompatibility. We told the truth to the family and got discharged for the effort, though we found later, through a relative, that telling the truth had good effect. In both the child and the parents the endocrines were constantly abused and with terrific results. Many a child is "Called Home" under such treatment.

Coffee and teas in excess disturb the endocrine function in a similar way to alcohol and tobacco. Many a coffee drinker has to have the stimulation of a cup of coffee in the morning before the vital forces (the endocrines) wake up to work.

Thoughts are very vital things and have been amply proved to be of constructive or destructive influence upon the building and activating forces of man. Thoughts receive their impetus from the endocrines and again react upon the endocrines either for good or evil as the thoughts may be. The first great urge within man is for food. This thought remains dominant throughout life. This urge includes not only actual things man puts into his mouth and stomach but it includes the comforts of life—clothing, home, dainties and all the refinements of eating. Much of man's thinking is about the getting and enjoying of food. The next great urge is for sex, which includes the union of male and female and all love, parent love, love of companionship, love of art, and all emotional and physical activity. The Creator surely created or evolved man in this way, and the original intent must have been pure. We evolve fastest by giving service, by helping each other, and by expressing harmony, idealism, order, and beauty.

The Human Temple is sacred. We must learn to build well physically, mentally, emotionally and spiritually, and thus learn the Fatherhood of God, the brotherhood of man, the continuous life and law of action and reaction—completely expressed life.

Our endocrines are making our personality, and the development of our mental and spiritual expressions will make our individuality. Too much of our teaching, in every line, makes for suppression of individuality. The child at school must learn to master or memorize many thoughts of other men or teachers. In Sunday School or church the child, or even the adult, must accept and believe the things told by the teacher, the priest, or the preacher. In the factory the worker must do as instructed. All this kind of teaching makes automatons, not thinkers. INDIVIDUALITY WILL NOT BE DEVELOPED UNTIL THE PERSON CAN THINK. Many persons think they think, but they really only think the things they have memorized. Many educators, preachers, doctors, lawyers, and bankers can quote you the writings of great writers and teachers and authorities and seem highly educated, and yet they have never really had an original thought nor emanated an initial idea. We believe in much reading but only for new food for thought and individual expression in an individualistic way. No person is educated until he can think for himself. To have been through college is no indication that the person can think. Abraham Lincoln never went through college yet he was one of the world's foremost thinkers.

If much of the junk that is being poured into the heads of our pupils of the present day could be forgotten, and the pupils taught to develop their endocrine powers and taught to think along the lines of their natural bents, then we could get a race of thinkers. Our pupils are mostly poor imitators at present. Purity of purpose and expression of life will come with freedom and naturalness of expression. The limitations that we put on life distort and cripple the natural expressions.

Every man must learn to speak the language of his soul. This he will do only when he recognizes the forces that are building him, and becomes capable of interpreting the urges within him. It takes a big vision to become a fine man. The world needs men of big vision, men that are virile and have fine endocrines. Men that have creative force and can visualize and build for the future as well as for the present . . . men of master minds.

We can think of no finer reward for the effort of writing this book than the knowledge that someone was helped to a little finer thinking about his evolvement and a greater effort at self-expression, by reading what has been written here.

To evolve together harmoniously is joy enough for all.

CHAPTER IX

Helpful Items

BOYS and girls entering puberty need a plentiful supply of calcium and phosphorus to develop their endocrines, especially the sex, thyroid and pineal glands.

The "fair, fat and forty" women entering the menopause period, very often, and usually, need thyroid extract.

A scientific report has just been made stating that cows which have been fed extract of pituitary increased the milk flow from ten to three hundred and fifty per cent. Why not also the human?

We have long known that X rays have an especial affinity for the destruction of abnormal or fast-growing cells and germs, and are particularly destructive to germs in their mitosis, or cell division, stage. Now science further finds that X rays kill living cells by suffocation. The action of the X rays on the tissues destroys the oxygen-getting power of the cells and so they die.

The X ray is becoming a great aid in stimulating or inhibiting the endocrines and especially the lymphatic ducts and glands, in certain types of persons.

Ulcers of the stomach will yield more quickly to the change in divine vibrations than to any other method of treatment. The author has proved this many times in his daily practice.

All living humans must have air in their stomachs and intestines. Too much quack advertising has made the race flatus conscious. It takes from 48 to 50 hours for the food to pass from the mouth to the rectum. Constant physic does not give the food time to digest and absorb. Give the bowels a chance to do their work. In most cases the endocrines with their auxiliary ducts and glands will do their work if they are left alone and have simple food to act upon.

If a man becomes irritable towards middle age, despondent and suicidal, and can see only the dark side of life, he should be very carefully examined for prostate trouble. Many a man has committed suicide just because he had an inflamed or enlarged prostate.

If a woman is emotional and cries easily and is despondent and even suicidal, she has an inflamed or diseased cervix of the womb and especially of the inner neck of the womb. This condition may have been brought on by bad care at childbirth, or she may have colon bacilli infection. Or she may be the victim of her husband's dose of gonorrhea that he had when a boy or before he was married. Most slaughtering of the ovaries and tubes comes from gonorrheal infection. This destruction of the female sex organs must be stopped and will be when men and women are properly examined before marriage and the public knows the full truth of the sources of gonorrhea and syphilis.

Both long and short wave diathermy when properly used are great factors in curing infections in the female and male sex organs.

Normally healthy sex activities build up a man and a woman and are their idealization, their imaging and creative power.

The man or woman with normal endocrines will be able to resist most diseases.

Sinus and antrum troubles are being absolutely cured by long and short wave diathermy. Our best results have come from the use of the old autocondensation current.

We are built through the forces of Light, Heat, Moisture,

and Movement. Should we not then study these forces in relation to our well-being? All these forces have direct bearing on the endocrines. The endocrines make us what we are. The understanding of helio- and electro-therapy has been a great help and still needs deeper study.

The highest life-growing foods are those that grow on trees or bushes.

Second best, those that grow above the ground as leafy vegetables, celery, lettuce, spinach, etc.

Third best, the underground or roots as potatoes, carrots, turnips, etc.

The best food can be made useless by bad cooking or bad preparation.

Milk and cheese contain large quantities of calcium and phosphates. Liver contains copper and iron. Eggs contain sulphur. Cod-liver oil and shellfish contain iodine. Vegetables contain cellulose as it is their cell membrane. These are very necessary for cell building, even as necessary as the vitamins of food.

Vitamin A is an anti-infectious substance. The condition is generally low where it is needed. It seems to have an especial affinity for eye, lung, sinus, and skin diseases. Vitamin A is found in cod-liver oil, yellow foods, milk, butter, yellow of eggs.

Vitamin B is antineurotic and is required in gastric intestinal diseases, in neuralgia, neuritis and constipation. Vitamin B is found in Brewer's yeast, germs of wheat, and vegetables cooked without salt. Salt may be added after cooking.

Vitamin C is antiscorbutic (scurvy tendency). It is needed in proper bone and teeth maintenance. Vitamin C is found in alkaline vegetables, lemons, and oranges, grapefruit and pineapples.

Vitamin D is antirachitic. It regulates the mineral metabolism, the bone-forming elements, calcium and phosphates. This is very necessary to the pregnant mother that she may have a perfect supply of calcium and phosphates. The lack of Vitamin D produces rickets. Vitamin D is found in cod-liver oil, milk, butter, yellow of eggs, and by body chemistry of the sun or ultraviolet rays. Persons subject to arthritis should beware of large doses of Vitamin D.

Vitamin E is necessary in the reproductive functions of the body. Vitamin E is found in the germs of wheat.

All diseases have vitamin and endocrine deficiency.

Sanitation goes hand in hand with civilization.

One of the greatest aids in sanitation is the lowly "patent toilet." In recent unearthings in Mesopotamia, it was found that the people of that age dumped their offal and refuse and slops just outside of the tent or house and as the space became filled they shoved it under their abode and started all over again. Think of the stenches and the diseases from such methods. Yet to this day peoples are still living like the ancient Mesopotamians. Sanitation is advancing quite rapidly and the patent toilet is no longer just for a city dweller. Farming communities are installing cesspools and the patent toilet.

If you are not feeling in normal health consult a doctor—one that will teach you about your condition. The word *doctor* means *teacher*. Every case is a law unto itself. Every case must be studied separately.

Health is not something that can be handed to you in a pill or in a surgical operation or in a light or electrical treatment. All these may help if intelligently used. The cause and effects of disease must be carefully studied. Read and reread this book and you will find great help in understanding your life forces.

You and all humanity can be uplifted by studying the creative and spiritual forces within yourself.

Humanity never has been and never will be uplifted by stepping over the dead and mutilated bodies of children, women and the flower of our manhood. Humanity must study and know the higher forces, within, that are related to the God or Creative forces from without.

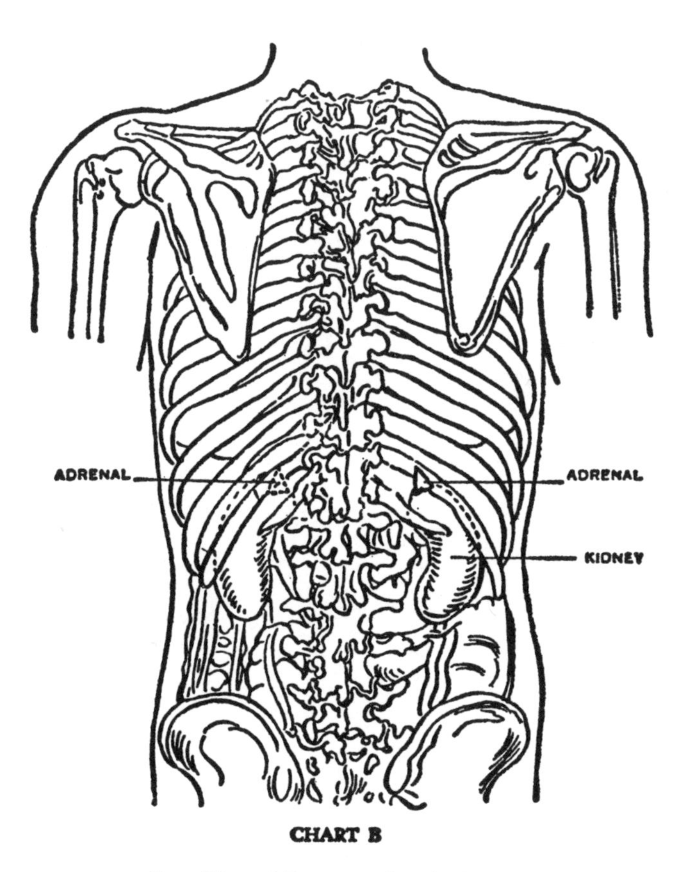

CHART B

Rear View of Location of Body Organs

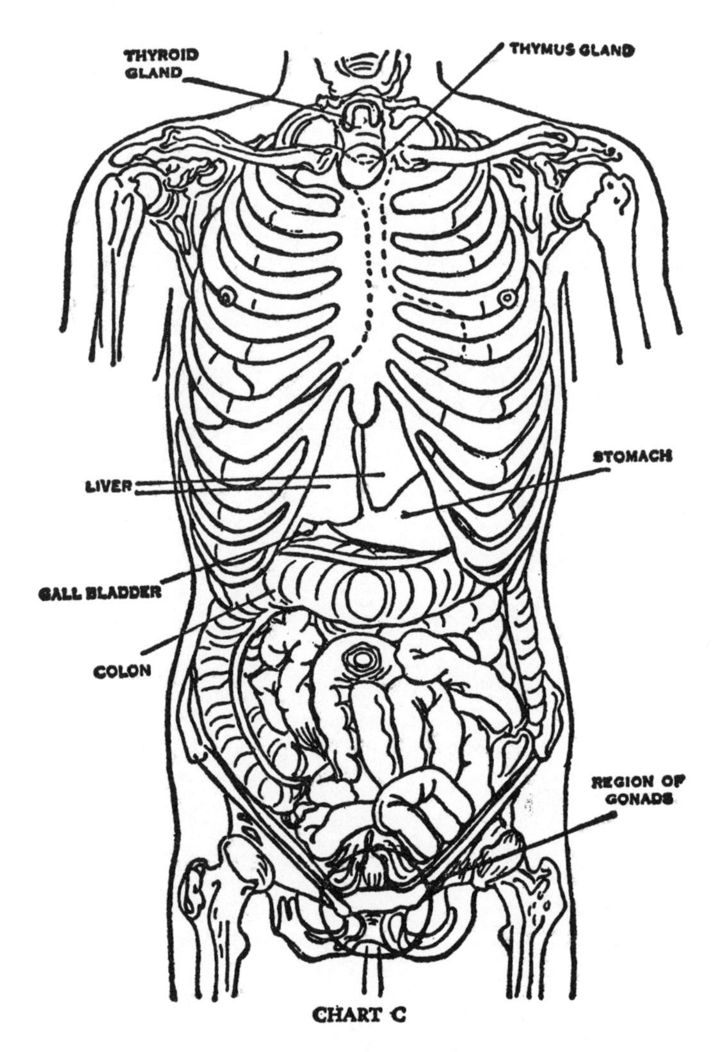

CHART C

Front View of Location of Body Organs

INDEX

▽

INDEX—*Continued*

INDEX—*Continued*

▽ ▽ ▽

Explanatory

∇

THE ROSICRUCIAN ORDER

ANTICIPATING questions which may be asked by the readers of this book, the publishers wish to announce that there is but one universal Rosicrucian Order existing in the world today, united in its various jurisdictions, and having one Supreme Council in accordance with the original plan of the ancient Rosicrucian manifestoes.

This international organization retains the ancient traditions, teachings, principles, and practical helpfulness of the Brotherhood as founded centuries ago. It is known as the *Ancient Mystical Order Rosae Crucis,* which name, for popular use, is abbreviated into AMORC. The headquarters of the Worldwide Jurisdiction is located at San Jose, California.

Those interested in knowing more of the history and present-day helpful offerings of the Rosicrucians may have a *free* copy of the book entitled, *The Mastery of Life,* by sending a definite request to SCRIBE G. K. A., Rosicrucian Park, San Jose, California 95191.

The Rosicrucian Library

Consists of a number of unique books which are described in the following pages.

ROSICRUCIAN SUPPLY BUREAU

ROSICRUCIAN PARK, SAN JOSE, CALIFORNIA 95191, U.S.A.

Eternal Fruits of Knowledge

By CECIL A. POOLE, F.R.C.

TRUTHS are those thoughts which have a *continuous value* to man in inspiration and service. Down through the ages have descended the illuminating ideas of philosophers, mystics, and profound thinkers that are as realistic today as when conceived centuries ago. It has been rightly said that we stand upon the shoulders of those who have gone before.

Unfortunately, however, we often are not aware of the knowledge that has stood the test of time. Such knowledge can serve *you* as well in our time as it did men of the past. There are points of experience and understanding which are ageless in their benefit to mankind. What these golden gems of wisdom are, this book reveals clearly, concisely, and interestingly.

This volume deals with such subjects as the nature of the Absolute; Body, Mind, and Soul; Good and Evil; Human and Universal Purpose; and many other interesting topics. It is a well-printed paperbound book.

▽

Rosicrucian Questions and Answers with Complete History of the Order

By H. SPENCER LEWIS, F.R.C., Ph.D.

THIS volume contains the first complete, authentic history of the Rosicrucian Order from ancient times to the present day. The history is divided into two sections, dealing with the traditional facts and the established historical facts, and is replete with interesting stories of romance, mystery, and alluring incidents.

This book is a valuable one since it is a constant reference and guidebook. Questions that arise in your mind regarding many mystical and occult subjects are answered in this volume.

For many centuries the strange mysterious records of the Rosicrucians were closed against any eyes but those of the high initiates. Even editors of great encyclopedias were unable to secure the strange, fascinating facts of the Rosicrucian activities in all parts of the world. Now the whole story is outlined and it reads like a story from the land of the "Arabian Nights."

The book outlines answers to scores of questions dealing with the history, work, teachings, benefits, and purposes of the Rosicrucian fraternity. It is printed on fine paper, and indexed.

Rosicrucian Principles for the Home and Business

By H. SPENCER LEWIS, F.R.C., Ph.D.

THIS volume contains such principles of practical Rosicrucian teachings as are applicable to the solution of everyday problems of life, in business and in the affairs of the home. It deals exhaustively with the prevention of ill-health, the curing of many common ailments, and the attainment of peace and happiness, as well as the building up of the affairs of life that deal with financial conditions. The book is filled with hundreds of practical points dealing especially with the problems of the average businessman or person in business employ. It points out the wrong and right way for the use of metaphysical and mystical principles in attracting business, increasing one's income, promoting business propositions, starting and bringing into realization new plans and ideals, and the attainment of the highest ambitions in life.

Rosicrucian Principles for the Home and Business is not theoretical but strictly practical. It has had a wide circulation and universal endorsement not only among members of the organization, who have voluntarily stated that they have greatly improved their lives through the application of its suggestions, but among thousands of persons outside of the organization. It has also been endorsed by business organizations and business authorities.

The book is of standard size and indexed.

▽

The Mystical Life of Jesus

By H. SPENCER LEWIS, F.R.C., Ph.D.

THIS is the book that thousands have been waiting for—the real Jesus revealed at last! It was in preparation for a number of years and required a visit to Palestine and Egypt to secure a verification of the strange facts contained in the ancient Rosicrucian and Essene records.

It is a full account of the birth, youth, early manhood, and later periods of Jesus' life, containing the story of his activities in the times not mentioned in the Gospel accounts. The facts relating to the Immaculate Conception, the birth, Crucifixion, Resurrection, and Ascension will astound and inspire you. The book contains many mystical symbols, fully explained, original photographs, and an unusual portrait of Jesus.

Here is a book that will inspire, instruct, and guide every student of mysticism and religion. It is one of the most talked-about books ever written on the subject. Read it and be prepared for the discussions of it that you will hear among men and women of learning. Indexed for quick reference.

▽

The Secret Doctrines of Jesus

By H. Spencer Lewis, F.R.C., Ph.D.

Does the Bible actually contain the unadulterated words of Jesus the Christ? Do you know that from A.D. 328 until A.D. 1870, there were held twenty ecclesiastic or church council meetings in which *man* alone decided upon the content of the Bible? Self-appointed judges in these councils decided to expurgate the Bible, removing those sacred writings which did not please them. The Great Master's *personal* doctrines, of the utmost, vital importance to every man and woman, were buried in unexplained passages and parables. *The Secret Doctrines of Jesus,* by Dr. H. Spencer Lewis, eminent author of *The Mystical Life of Jesus,* for the first time *reveals* these *hidden truths.* Startling, fascinating, this book should be in every thinker's hands. It is beautifully bound and illustrated.

▽

"Unto Thee I Grant . . ."

By Sri. Ramatherio

This is one of the rarest Oriental mystery books known. It was translated by special permission of the Grand Lama and Disciples of the Sacred College in the Grand Temple in Tibet.

Here is a book that was written two thousand years ago, but was hidden in manuscript form from the eyes of the world and given only to the initiates of the temples in Tibet to study privately.

Out of the mystery of the past comes this antique book containing the rarest writings and teachings known to man with the exception of the Bible. Hundreds of books have been written about the teachings and practices of the *Masters of the Far East* and the adepts of Tibet, but none of them has ever contained the secret teachings found in this book.

The book deals with man's passions, desires, weaknesses, sins, strengths, fortitudes, ambitions, and hopes. It contains also the strange mystic story of the expedition into Tibet to secure this marvelous manuscript.

▽

A Thousand Years of Yesterdays

By H. SPENCER LEWIS, F.R.C., Ph.D.

HERE is a book that will tell you about the real facts of *reincarnation.* It is a story of the soul, and explains in detail how the soul enters the body and how it leaves it, where it goes, and when it comes back to Earth again, and why.

The story is not just a piece of fiction, but a *revelation of the mystic laws* and principles known to the Masters of the Far East and the Orient for many centuries, and never put into book form as a story before this book was printed. That is why the book has been translated into so many languages and endorsed by the mystics and adepts of India, Persia, Egypt, and Tibet.

Fascinating—Alluring—Instructive

Those who have read this book say that they were unable to leave it without finishing it at one sitting. The story reveals the mystic principles taught by the Rosicrucians in regard to reincarnation as well as the spiritual laws of the soul and the incarnations of the soul.

An attractively bound book, worthy of a place in anyone's library.

▽

Self Mastery and Fate with the Cycles of Life

By H. SPENCER LEWIS, F.R.C., Ph.D.

THIS book is entirely different from any other book ever issued in America, dealing with the secret periods in the life of each man and woman wherein the cosmic forces affect our daily affairs.

The book reveals how we may take advantage of certain periods to bring success, happiness, health, and prosperity into our lives, and it likewise points out those periods which are not favorable for many of the things we try to accomplish. It does not deal with astrology or any system of fortune-telling, but presents a system long used by the Master Mystics in Oriental lands and which is strictly scientific and demonstrable. One reading of the book with its charts and tables will enable the reader to see the course of his life at a glance. It helps everyone to eliminate "chance" and "luck," to cast aside "fate," and replace these with self mastery.

Here is a book you will use weekly to guide your affairs throughout the years. There is no magic in its system, but it opens a vista of the life-cycles of each being in a remarkable manner. This book is beautifully bound.

▽

Rosicrucian Manual

By H. SPENCER LEWIS, F.R.C., Ph.D.

THIS practical book contains not only extracts from the Constitution of the Rosicrucian Order, but a complete outline and explanation of all the customs, habits, and terminology of the Rosicrucians, with diagrams and explanations of the symbols used in the teachings, an outline of the subjects taught, a dictionary of the terms, a complete presentation of the principles of Cosmic Consciousness, and biographical sketches of important individuals connected with the work. There are also special articles on the Great White Lodge and its existence, how to attain psychic illumination, the Rosicrucian Code of Life with twenty-nine laws and regulations, and a number of portraits of prominent mystics including Master K.H., the Illustrious.

The technical matter in the text and in the numerous diagrams makes this book a real encyclopedia of Rosicrucian explanations, aside from the dictionary of Rosicrucian terms.

The *Rosicrucian Manual* has been enlarged and improved since its first edition. Attractively bound, and stamped in gold.

▽

Mystics at Prayer

Compiled by MANY CIHLAR

THE first compilation of the famous prayers of the renowned mystics and adepts of all ages.

The book *Mystics at Prayer* explains in simple language the reason for prayer, how to pray, and the cosmic laws involved. You come to learn the real efficacy of prayer and its full beauty dawns upon you. Whatever your religious beliefs, this book makes your prayers the application not of words, but of helpful, divine principles. You will learn the infinite power of prayer. Prayer is man's rightful heritage. It is the direct means of man's communion with the infinite force of divinity.

▽

Behold the Sign

By Ralph M. Lewis, F.R.C.

What were the *Sacred Traditions* said to have been revealed to Moses—and never spoken by the ancient Hebrews? What were the forces of nature discovered by the Egyptian priesthood and embodied in strange symbols—symbols which became the everliving knowledge which built King Solomon's Temple, and which found their way into the secret teachings of every century?

Regardless of the changing consciousness of man, certain signs and devices have immortalized for all ages the truths which make men free. Learn the meaning of the Anchor and Ark, the Seven-Pointed Star, ancient Egyptian hieroglyphs, and *many other age-old secret symbols.*

Here is a book that also explains the origin of the various forms of the cross, the meanings of which are often misunderstood. It further points out the mystical beginnings of the *secret signs* used by many fraternal orders today. This book of symbolism is *fully illustrated,* simply and interestingly written.

▽

Mansions of the Soul

By H. Spencer Lewis, F.R.C., Ph.D.

Reincarnation! The world's most disputed doctrine. The belief in reincarnation has had millions of intelligent, learned, and tolerant followers throughout the ages. Ringing through the minds and hearts of students, mystics, and thinkers have always been the words: "Why Are We Here?" Reincarnation has been criticized by some as conflicting with sacred literature and as being without verification. This book reveals, however, in an intelligent manner the many facts to support reincarnation. Quotations from eminent authorities, and from Biblical and sacred works substantiate reincarnation. This volume *PROVES* reincarnation, placing it high above mere speculation. Without exaggeration, this is the most complete, inspiring, and enlightening book ever written on this subject. It is not fiction but a step-by-step revelation of profound mystical laws. Look at *some* of the thought-provoking, intriguing chapters:

The Personality of the Soul; Does Personality Survive Transition?; Heredity and Inheritance; Karma and Personal Evolution; Religious and Biblical Viewpoints; Christian References; Souls of Animals and the "Unborn"; Recollections of the Past.

Over 300 pages. Beautifully printed, neatly bound, stamped in gold —a valuable asset to your library.

▽

Lemuria—the Lost Continent of the Pacific

By WISHAR S. CERVÉ

BENEATH the rolling, restless seas lie the mysteries of forgotten civilizations. Swept by the tides, half-buried in the sands, worn away by terrific pressure, are the remnants of a culture little known to our age of today. Where the mighty Pacific now rolls in a majestic sweep of thousands of miles, there was once a vast continent. This land was known as Lemuria, and its people as Lemurians.

We pride ourselves upon the inventions, conveniences, and developments of today. We call them modern, but these ancient and long-forgotten people excelled us. Things we speak of as future possibilities, they knew as everyday realities. Science has gradually pieced together the evidences of this lost race, and in this book you will find the most amazing, enthralling revelations you have ever read. How these people came to be swept from the face of the Earth, except for survivors who have living descendants today, is explained. Illustrations and explanations of their mystic symbols, maps of the continent, and many ancient truths and laws are contained in this unusual book.

If you are a lover of mystery, of the unknown, the weird—read this book. Remember, however, this book is *not fiction,* but based on facts, the result of extensive research. Does civilization reach a certain height and then retrograde? Are the culture and progress of mankind in cycles, reaching certain peaks, and then returning to start over again? These questions and many more are answered in this intriguing volume. Read of the living descendants of these people, whose expansive nation now lies within the Pacific. These descendants have the knowledge of the principles which in bygone centuries made their forebears builders of an astounding civilization.

The book, *Lemuria—the Lost Continent of the Pacific,* is beautifully bound, well printed, and contains many illustrations.

Whisperings of Self

By VALIDIVAR

Whisperings of Self is the interpretation of cosmic impulses received by a great mystic-philosopher, Ralph M. Lewis, who in this work writes under thc pen name of Validivar.

The aphorisms in this collection have appeared singly in copies of the *Rosicrucian Digest* over a period of forty years and comprise insights into all areas of human experience—justice, war and peace, ethics, morals, marriage, family, work, leisure, and countless others.

Ralph Lewis' frank and direct style provides much food for thought in each brief impression. A reader develops the habit of using a thought a day, and there are more than two hundred from which to choose.

This is an attractive, hardcover book that makes an attractive gift as well as a treasured possession of your own.

▽

The Symbolic Prophecy of The Great Pyramid

By H. SPENCER LEWIS, F.R.C., Ph.D.

THE world's greatest mystery and first wonder is the Great Pyramid. It stands as a monument to the learning and achievements of the ancients. For centuries its secrets were closeted in stone—now they stand revealed.

Never before in a book priced within the reach of every reader have the history, vast wisdom, and prophecies of the Great Pyramid been given. You will be amazed at the Pyramid's scientific construction and at the tremendous knowledge of its mysterious builders.

Who built the Great Pyramid? Why were its builders inspired to reveal to posterity the events of the future? What is the path that the Great Pyramid indicates lies before mankind? Within the pages of this enlightening book there are the answers to many enthralling questions. It prophesied the World Wars and the great economic upheaval. Learn what it presages for the future. You must not deprive yourself of this book.

The book is neatly and attractively bound, and contains instructive charts and illustrations.

The Book of Jasher

THE SACRED BOOK WITHHELD

BY WHAT right has man been denied the words of the prophets? Who dared expunge from the Holy Bible one of its inspired messages? For centuries man has labored under the illusion that there have been preserved for him the collected books of the great teachers and disciples—yet one has been withheld—The Book of Jasher, discovered by Alcuin in A.D. 800. Later it was suppressed, then rediscovered in 1829, and once again suppressed.

Within the hallowed pages of the great Bible itself are references to this lost book. As if by divine decree, the Bible appears to cry out to mankind that its sanctity has been violated, its truth veiled, for we find these two passages exclaiming: "Is not this written in the Book of Jasher?"—Joshua 10:13; "Behold, it is written in the book of Jasher" —2 Sam. 1:18.

An actual photographic reproduction of this magnificent work, page for page, line for line, unexpurgated. Bound in its original style.

▽

Herbalism Through the Ages

By RALPH WHITESIDE KERR, F.R.C.

VERY FEW things in human experience have touched the whole being of man as have herbs. Not only did they provide man's earliest foods and become remedies and medicines for his illnesses, but they also symbolized certain of his emotions and psychic feelings. Further, herbs are one of Nature's products that we still depend upon for their virtues, even in our modern age.

The source of our first foods has a romantic and fascinating history. This book reveals man's discovery of natural foods, herbs, and their various uses through the centuries. Most medicines prescribed or purchased today owe their healing or pain-relieving value to the properties of herbs or herbal products. Certain herbs are a natural medicine; they have a health-giving essence. Modern medical science uses many herbs whose real identity is obscured by technical medical terms. This book lists many of these herbs and tells their history and use.

Mental Poisoning

THOUGHTS THAT ENSLAVE MINDS

By H. SPENCER LEWIS, F.R.C., Ph.D.

TORTURED souls. Human beings, whose self-confidence and peace of mind have been torn to shreds by invisible darts—thc evil thoughts of others. Can envy, hate, and jealousy be projected through space from the mind of another? Do poisoned thoughts like mysterious rays reach through the ethereal realm to claim innocent victims? Will wishes and commands born in hate gather momentum and like an avalanche descend upon a helpless man or woman in a series of calamities? Must humanity remain at the mercy of evil influences created in the minds of the vicious? Millions each year are mentally poisoned—are you safe from this scourge? *Mental Poisoning* fearlessly discloses this psychological problem. Read its revelations and be prepared.

This neatly bound, well-printed book has been economically produced so it can be in the hands of thousands because of the benefit it will afford them.

▽

Glands—Our Invisible Guardians

By M. W. KAPP, M.D.

YOU NEED not continue to be bound by those glandular characteristics of your life which do not please you. These influences, through the findings of science and the mystical principles of nature, may be adjusted. The first essential is that of the old adage: "Know Yourself." Have revealed the facts about the endocrine glands—know where they are located in your body and what mental and physical functions they control. The control of the glands can mean the control of your life. These facts, scientifically correct, with their mystical interpretation, are presented in simple, nontechnical language, which everyone can enjoy and profit by reading.

Mystics and metaphysicians have long recognized that certain influences and powers of a cosmic nature could be tapped; that a divine energy could be drawn upon, affecting our creative ability, personality, and our physical welfare. For centuries there has been speculation as to what area or what organs of the body contain this medium—this contact between the Divine and the physical. Now it is known that certain of the glands are governors which speed up or slow down the influx of cosmic energy into the body. This process of divine alchemy and how it works is explained in this book of startling facts.

Dr. M. W. Kapp, the author, during his lifetime, was held in high esteem by the medical fraternity, despite the fact that he also expressed a deep insight into the mystical laws of life and their influence on the physical functioning of the body.

INTRODUCTION BY H. SPENCER LEWIS, F.R.C., Ph.D.

Dr. H. Spencer Lewis—former Imperator of the Rosicrucian Order (AMORC), for its present cycle of activity, and author of many works on mysticism, philosophy, and metaphysics—wrote an important introduction to this book, in which he highly praised it and its author.

▽

What to Eat—and When

By Stanley K. Clark, M.D., C.M., F.R.C.

"Mind over matter" is not a trite phrase. Your moods, your temperament, your very *thoughts* can and *do* affect digestion. Are you overweight—or underweight? Appearances, even the scales, are not always reliable. Your age, your sex, the kind of work you do—all these factors determine whether your weight is correct or wrong for *you.* Do you know that many people suffer from food allergy? Learn how your digestion may be affected even hours after you have eaten.

The author of this book, Dr. Stanley K. Clark, was for several years staff physician at the Henry Ford Hospital in Detroit. He is a noted gastroenterologist (specialist in stomach and intestinal disorders). He brings you his wealth of knowledge in this field, *plus* his additional findings from his study of the effects of the *mind* upon digestion.

What to Eat—and When is compact, free from unnecessary technical terminology. Includes complete handy index, *food chart,* and *sample menus.* It is not a one-time-reading book. You will often refer to it throughout the years. Well printed, strongly bound.

Sepher Yezirah—A Book on Creation OR THE JEWISH METAPHYSICS OF REMOTE ANTIQUITY

DR. ISIDOR KALISCH, Translator

AMONG the list of the hundred best books in the world, one might easily include this simple volume, revealing the greatest authentic study of the secret Kabala. For those averse to fantastic claims, this book is truly *comprehensible*—for the wise student who does not care for magical mumbo-jumbo, it is *dynamic*.

The phantasies of those baffling speculations of other writers become unimportant when the practical student of mysticism reverently thumbs through these pages and catches the terse and challenging statements. The woolgathering of many so-called authors of occultism is brought to nothing by this simple volume which makes a pattern for honest mystical common sense.

The *Sepher Yezirah* contains 61 pages with both Hebrew and English texts, photolithographed from the 1877 original edition. For anyone interested in the best—also, considered by some, the most ancient—in Hebrew mystical thought, this book will be a refreshing discovery.

The careful reader will be attracted to three characteristics of this edition of the *Sepher Yezirah:*

(1) A clear English translation of a most ancient work, almost unavailable up to the present.
(2) A simple exposé of fundamental aspects of the ancient Kabala without superstitious interpretations.
(3) An inexpensive translation of the world's oldest philosophical writing in Hebrew.

Attractive and convenient, paperbound edition.

The Conscious Interlude

By RALPH M. LEWIS, F.R.C.

HOW MANY of the countless subjects which shape your life are inherited ideas? How many are actually yours? Would you like to have your own mind look at itself in perspective for an analysis? In this book, Mr. Lewis, Imperator of the worldwide Rosicrucian Order, AMORC, outlines the culmination of years of his original thought. As you follow him through the pages into broad universal concepts, your mind too will feel its release into an expanding consciousness.

You will be confronted with and will answer such questions as: Is consciousness something innate or is it generated? What is the reality that you experience *actually like?* What are your own conscious interludes? This work belongs to every seeker after knowledge. Indexed and illustrated, this is a volume of more than 360 pages.

▽

Essays of A Modern Mystic

By H. SPENCER LEWIS, F.R.C., Ph.D.

THE writings of a true mystic philosopher constitute cosmic literature. The ideas they contain are born of *inner experience*—the self's contact with the cosmic intelligence residing within. Such writings, therefore, have the ring of conviction—of *truth.*

This book, *Essays of A Modern Mystic,* will disclose the personal confidence and enlightenment that mystical insight can give an individual.

The essays are a compilation of the private writings by Dr. H. Spencer Lewis which have never before been published in book form. Dr. Lewis is not only the author of many literary works but also was a contributor to publications and periodicals with worldwide circulation.

This book is not hastily put together. It has a hard binding, attractively stamped in gold.

▽

Cosmic Mission Fulfilled

By RALPH M. LEWIS, F.R.C.

THE life of Harvey Spencer Lewis, Imperator of the Ancient, Mystical Order Rosae Crucis, is a fascinating account of the struggle of a mystic-philosopher against forces of materialism. He was charged with the responsibility of rekindling the ancient flame of wisdom in the Western world.

In the life of this great man events swung like a pendulum from triumph to tribulation. These became a progressive stimulus to achievement.

Has each man a purpose on the earth plane? Our happiness lies in understanding this and in the realization of hopes worthy of our best personal powers. The present is our *moment in Eternity*. In it we fulfill our mission.

In this book, the author combines his close and official knowledge of Dr. Lewis with the anecdotes of many other persons who knew him.

Nine full-plate illustrations are inserted into this beautifully printed and bound volume.

▽

Egypt's Ancient Heritage

By RODMAN R. CLAYSON

MUCH of what we know today began in Egypt! Out of that ancient civilization, which lasted three thousand years, came the first concepts of the origin of the universe, clothed in a symbology that showed a marvelous insight into natural law. Truth, righteousness, justice, and moral codes were first taught in the ancient mystery schools of Egypt as was the belief in a mind cause or thought as the creative cosmic force.

The belief in the soul, of life after death, and of immortality was held by the Egyptians thousands of years before Christ. The judgment of the soul in the next life evolved from an Egyptian concept and was dramatized by their rites and ceremonies.

This book tells of the amazing similarity of Egyptian thought to modern religious, mystical, and philosophical doctrines, and how many of our customs and beliefs of today were influenced by these ancient people. It is truly an amazing revelation!

Written in a straightforward, easily read style, *Egypt's Ancient Heritage* is hardbound.

The Sanctuary of Self

By RALPH M. LEWIS, F.R.C.

WHAT could be more essential than the discovery and analysis of *self,* the composite of that consciousness which constitutes one's whole being? This book of sound logic presents revealingly and in entirety the four phases of human living: The Mysteries, The Technique, The Pitfalls, and Attainment.

Do you not, at times, entertain the question as to whether you are living your life to your best advantage? You may find an answer in some of the 23 chapters, presented under headings such as: Causality and Karma, The Lost Word, Death: The Law of Change, Love and Desire, Nature of Dreams, Prediction, Mastership and Perfection. Consider "Love and Desire." In much of ancient and modern literature, as well as in the many and various preachments of the present-day world, LOVE is proclaimed as the solution to all human conflict. Do you understand truly the meaning of *absolute love?* Do you know that there are various *loves* and that some of the so-called loves are dangerous drives?

Written authoritatively by Ralph M. Lewis, Imperator of the Rosicrucian Order (AMORC), this volume of over 350 pages, carefully indexed, is of particular value as a text for teachers and students of metaphysics, including philosophy and psychology.

▽

Yesterday Has Much to Tell

By RALPH M. LEWIS, F.R.C.

MAN'S conquest of nature and his conflict with self, as written in the ruins of ancient civilizations, found in the sacred writings of temples and sanctuaries, and as portrayed in age-old tribal rites, are related to you by the author from his extensive travels and intimate experiences. This is not a mere travel book. It constitutes a personal witnessing and account of primitive ceremonies, conversations with mystical teachers and austere high priests of the Near and Far East. It takes you into the interior of Africa to see the performance of a witch doctor and to temples in Peru, India, Egypt, and other exotic lands. The author was privileged because of his Rosicrucian affiliation to see and to learn that which is not ordinarily revealed. A hardbound book of 435 pages, including sixteen pages of photographs.

The Technique Of The Master

THE WAY OF COSMIC PREPARATION

By RAYMUND ANDREA, F.R.C.

A GUIDE to inner unfoldment! The newest and simplest explanation for attaining the state of Cosmic Consciousness. To those who have felt the throb of a vital power within, and whose inner vision has at times glimpsed infinite peace and happiness, this book is offered. It converts the intangible whispers of self into forceful actions that bring real joys and accomplishments in life. It is a masterful work on psychic unfoldment.

Cares That Infest

By CECIL A. POOLE, F.R.C.

WE EACH have problems—but it is how we solve them that affects our individual development and our relationships with others. Learning comes both from our problems and from our solutions.

Realizing our weaknesses and basing our lives on a workable system of values will help each of us in our personal evolution. This book discusses such specific problems as: worry, fear, and insomnia, and the development of a practical philosophy of life to alleviate the suffering caused by these difficulties.

This volume is attractively printed, bound, and stamped in gold.

▽

Mental Alchemy

By RALPH M. LEWIS, F.R.C.

Now available—the latest volume by this distinguished and well-respected author of mystical, metaphysical, and practical philosophical literature.

Are we each responsible for the creation of our own surroundings?

Perhaps not entirely—but by the proper mental attitude we can alter certain aspects of our lives, making them more compatible with our goals. It is easier to cope with a difficulty if we realize that, to some extent, we can transmute the problem to a workable solution through *mental alchemy*. The process is neither easy nor instantaneously effective.

Long hours of thought, frank appraisals of our goals and the goals of others, and honest assessments of personal capabilities are necessary elements for the process of *mental alchemy*. Eventually, however, the serious person will be rewarded by having the ability to alter the direction of his life through proper thought and the understanding of the elements involved.

This volume is attractively printed, bound, and stamped in gold.

▽